DRAWING OF THE HAND

AND ITS ANATOMY

DRAWING

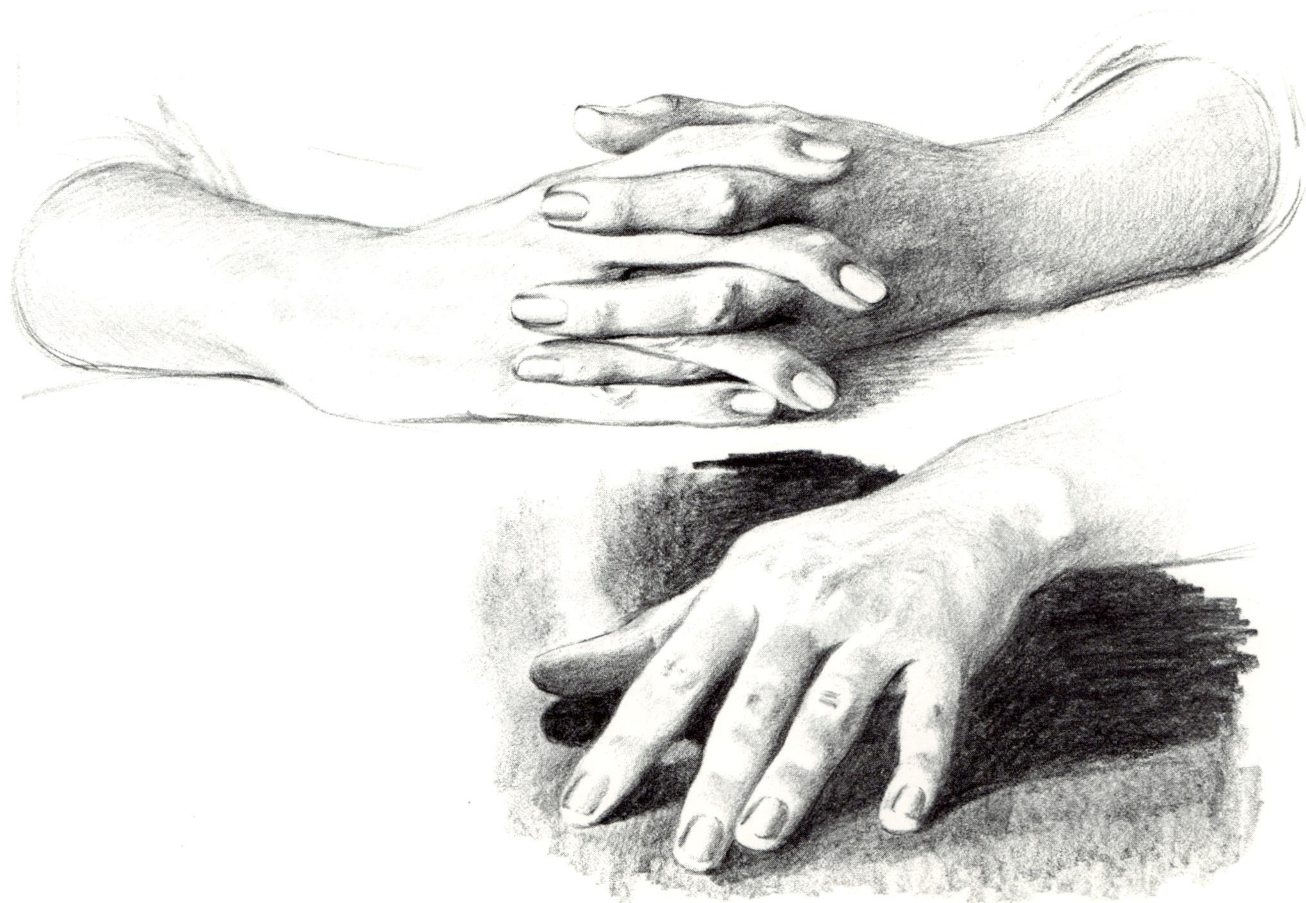

OF THE HAND

AND ITS ANATOMY

by Joseph M. Henninger

1973

BORDEN PUBLISHING COMPANY

ALHAMBRA, CALIFORNIA

First Editon

LIBRARY OF CONGRESS CATALOG CARD NUMBER: 73-77927

PRINTED IN THE UNITED STATES OF AMERICA

INTRODUCTION

MICHELANGELO'S REMARKS on drawing have often been quoted, and I am sure that most of the people who have ever had anything to do with the subject would agree that the possession of the ability to draw is truly, in his words, "a great treasure."

The possession of this is so difficult to achieve, however, that there are not many people today who have any desire to make the effort. One wonders, now that there is a strong resurgence of interest in drawing, just where the aspiring young draughtsman will be able to learn his craft, as there are very few teachers left who can help a student through the discipline and the rigors of experiencing form in the most direct way.

Joseph Morgan Henninger who made the drawings in this book is one of the few. His knowledge of drawing and anatomy is profound, as this book proves, and his experience with those subjects covers thirty-six years as a teacher and head of the Department of Illustration at Art Center College, in Los Angeles, California. During that time he was responsible for the development of literally thousands of devoted young students to a point where they were well equipped to go out and earn their living in the various branches of the graphic arts.

On his own account his career has been an outstanding one. At an early age he was awarded a scholarship to the Herron Art Institute in Indianapolis, Indiana, and he subsequently won a three year scholarship to the École des Beaux Arts, Paris, where he studied under Lucien Simon. After returning to this country as a qualified landscape and portrait painter, he married the brilliant young Sorbonne student he met on the *Leviathan* during the homeward voyage and eventually moved to Los Angeles. From then on, he progressed smoothly into advertising illustration and teaching, with time out during the war years for a job of drawing fighter planes for Lockheed. At the close of the war he resumed his twin careers, and in the advertising field was the winner of thirty-two awards, eight of which were medals. He was the first president of the Los Angeles Society of Illustrators and the only illustrator ever elected to the presidency of the Los Angeles Art Directors Club. He was chairman of the Art Advisory Board of the State of California for three years and acting president of the California Arts Club for two.

During all this time he was an active three-day-a-week teacher at Art Center and his classes consistently swept the major prizes in the national scholarship competitions. During the past ten years, as time allowed from his busy schedule, he has worked at compiling this monumental job which in my estimation is the best of all the books on the subject that have ever been published. It should guarantee, for all time, Joe Henninger's reputation as an outstanding expert in that field, and should also prove to be the practical reference book on hands that the artists, craftsmen and students have so long and so sorely needed.

Donald Teague, N.A.A.W.S.

AUTHOR'S NOTES

THIS BOOK has been written with three objectives in mind:

1. To fulfill a need for a book devoted exclusively to the drawing of the hand.
2. To clarify, as much as possible, the problems commonly associated with the drawing of the hand.
3. To lay at rest the old saw that "hands are impossible to draw."

After some thirty or more years as an instructor at Art Center College of Design, I have concluded that art students are not avid readers but are "picture lookers." Since this seems to be the norm, the text of this book shall be held to the minimum, and explanatory notes accompanying the plates shall be as minimal as clarity permits.

This book is not intended as a catalogue of gestures, neither is it intended as a copybook of drawing techniques. Hopefully, it shall be an inspiration to the serious student to acquire a greater knowledge, and achieve mastery in drawing the hand.

The plates in this book have been reproduced at the actual scale of the original drawings. The medium used in their execution is given on each plate.

Aspiring art students devote years of arduous study to the drawing of the head and figure, yet often times give short shrift to the study of the hand. Many a competent portrait has had its quality reduced by poorly painted hands. All manner of devices are devised to avoid the task of painting or drawing the hands. Illustrators, however, realize the importance the hand plays in fortifying the expression of emotion and are, by and large, excellent draughtsmen of the hand. Actors and dancers are aware of what part the proper gesture plays in conveying an emotion. The pantomimist's art relies heavily upon the hands to express his message. The hand itself without the spoken word becomes the spokesman, to wit: the raised clenched fist — defiance; the clenched fist hanging at the side — resolution; the pointed index finger — accusation; the cathedral like gesture of the two hands — prayer; the hands raised and open — supplication; and on and on, the list seems endless.

The hand assists in establishing character — to wit — the heavy, muscular hand of the laborer drawn on the arm of a young beautiful girl would be ridiculous and vice versa. The age and physical condition of the model is mirrored in the hands, ergo the hands are extremely important to arrive at total characterization.

Fortuitous evolutionary development of man's thumb, and its placement, has made man the dexterous creature he is. Many animals, starting with the great apes and the monkeys, have thumbs, but the placement of these digits renders them virtually useless for great dexterity. Only man can make his thumb approximate with each of his four fingers with ease. This happy ability has made him capable of executing the tasks that his fertile brain originates. He has built the most delicate mechanisms and constructed awe inspiring structures. Imagine, if you will, Edison's brain and the paw of a dog or the hoof of a horse at the end of his arm. The delicate sensitivity of the hand allows the blind to assemble the most complex devices, or to enchant us with music from the multiple keyboards of an organ.

Why does the aspiring draughtsman find difficulty in drawing the hand? There are, I believe, three basic reasons: first a lack of knowledge of its anatomical construction; second its multiple members, namely, the wrist, the palm, and the five digits; and lastly its ability to assume endless complex shapes. If one has at least basic knowledge of its anatomical make-up, and will reduce its involved shapes to simple geometric pattern, then the most detailed drawings are achievable. It is my hope that this book shall prove my claims, and shall entice the serious student into study that shall reward him or her with a feeling of accomplishment.

ANATOMY

MANY STUDENTS react to the word "anatomy" with the repugnance that the word "cancer" or "snake" engenders. It is my belief that this is due greatly to the Greek and Latin nomenclature of the muscles and bones. The Germans, French, Russians and the Spanish have all translated these Latin terms into their respective languages. Only the English speaking people have left them untranslated. No one, who speaks English, finds difficulty in referring to the major muscle of the upper arm as the "biceps" — this word comes from the Latin "bicaput," pronounced *bee kahput,* which means "two heads." Those students who have a good background in Latin do not seem to recoil so much from the study of anatomy. The Latin terminology used is highly descriptive of either shape or action. I have composed an index giving the Latin and its English translation. I hope this shall overcome much of the mystery which seems to surround the study of anatomy. It is not necessary to be conversant with anatomical nomenclature to draw the body and its various parts, however, it is helpful if one is compelled to talk about it knowledgeably.

The hand contains twenty seven bones; eight in the wrist, four in the palm, three in the thumb and twelve in the fingers. The number of muscles in the hand proper are twenty: six for the thumb, three for the little finger and eleven between the bones in the palm and back of the hand. The muscles which control the bending and straightening of the fingers and thumb are located in the forearm as are those which control the movement of the wrist. These muscles communicate their action via a group of long tendons. The mass of the six muscles of the thumb, which originate in the hand, is called the "Thenar Eminence", while the mass composed of the three muscles on the little finger side is called the "Hypothenar Eminence." Both of these masses are found on the palmarside. There are two fleshy pads on the inside of the thumb and three fleshy pads on each finger on the palm side. Due to the fact that the palm is longer than the back of the hand (see drawing on page 25) there are four fleshy pads visible on the index finger and the little finger when the hand is closed into a fist. This often unobserved change makes a tremendous difference in the feeling of the strength in the drawing of the fist (see drawing on page 30).

The mass that is formed by the eight bones of the wrist makes a universal-joint-like hinge between the forearm and the hand. This mass cannot be seen on the palm side and is only slightly evident in the back of the hand when it is fully flexed, namely, bent sharply toward the palm. However, these bones make a delicate change in the line from the forearm into the hand that gives conviction to its drawing. (See drawings on page 18).

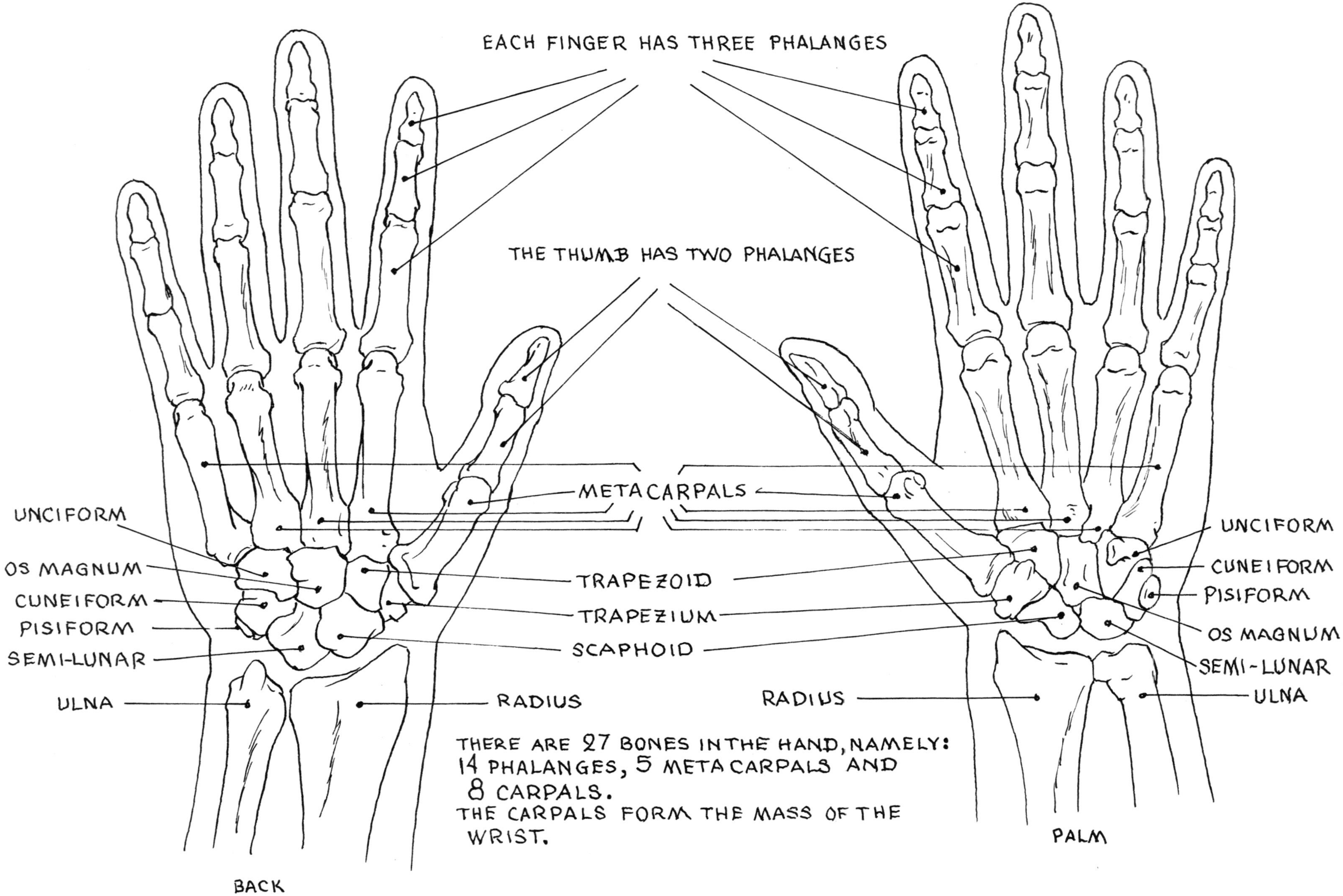

THERE ARE 27 BONES IN THE HAND, NAMELY: 14 PHALANGES, 5 METACARPALS AND 8 CARPALS.
THE CARPALS FORM THE MASS OF THE WRIST.

THE BONES OF THE WRIST (CARPALS).

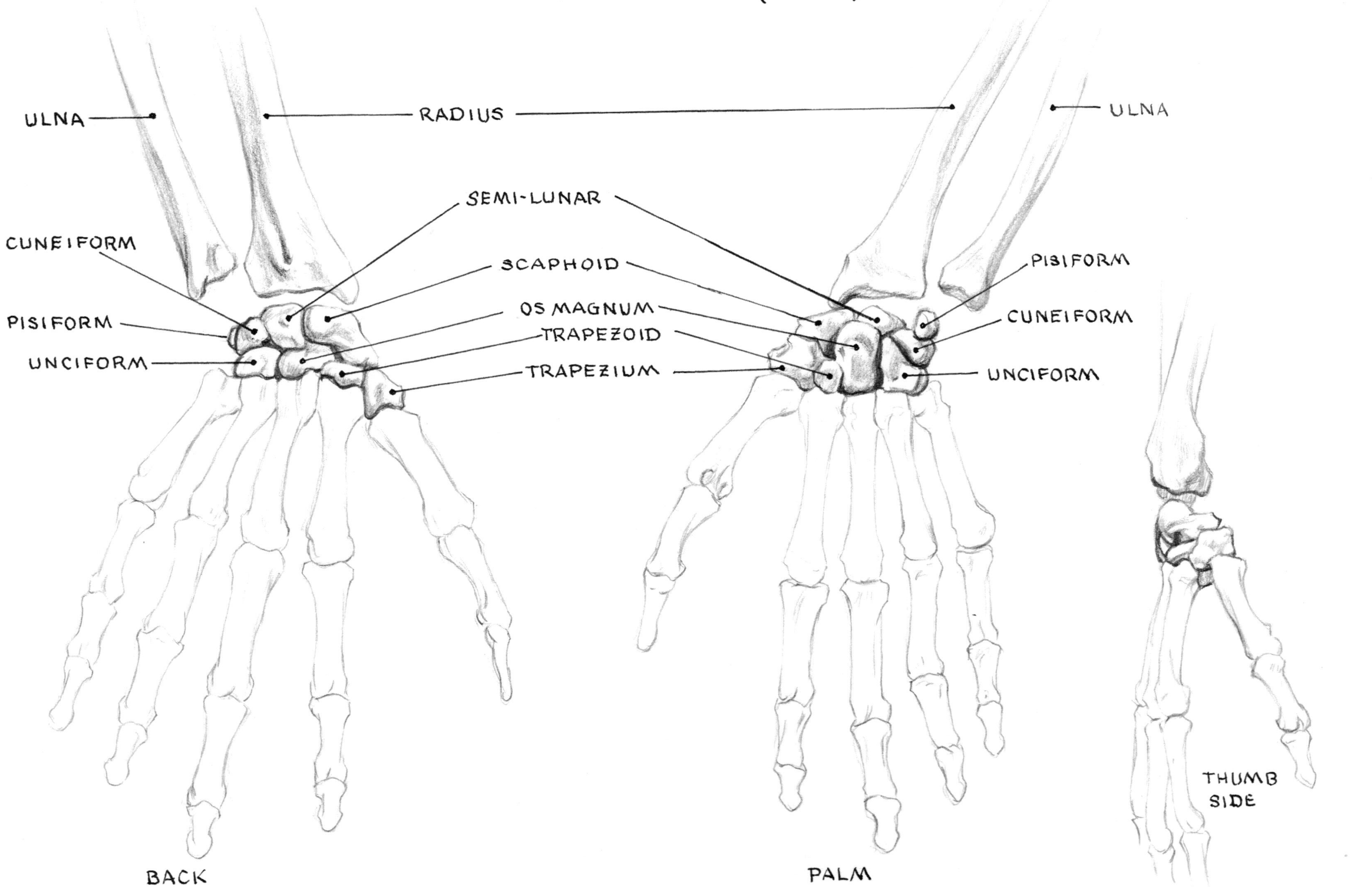

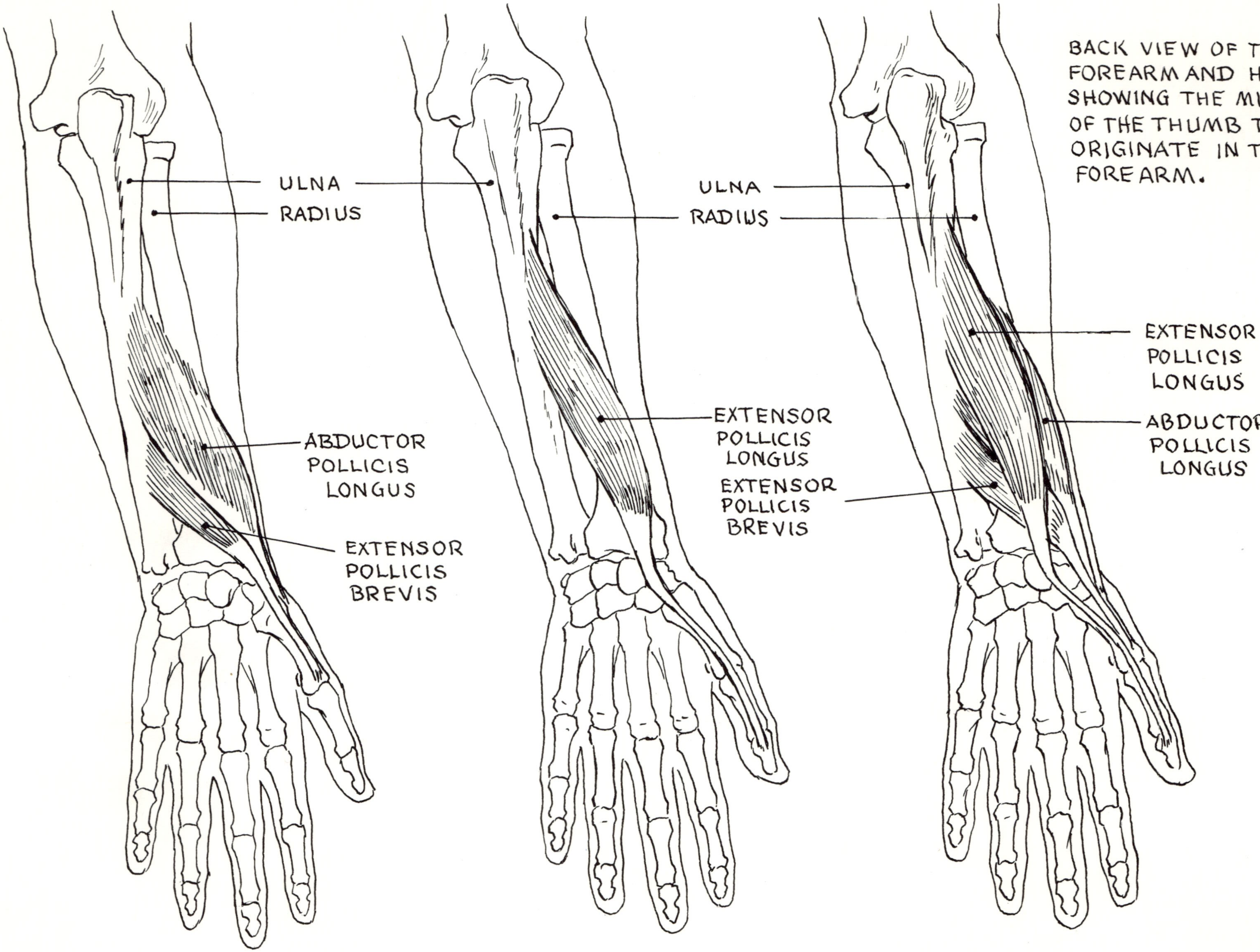

BACK VIEW OF THE FOREARM AND HAND SHOWING THE MUSCLES OF THE THUMB THAT ORIGINATE IN THE FOREARM.

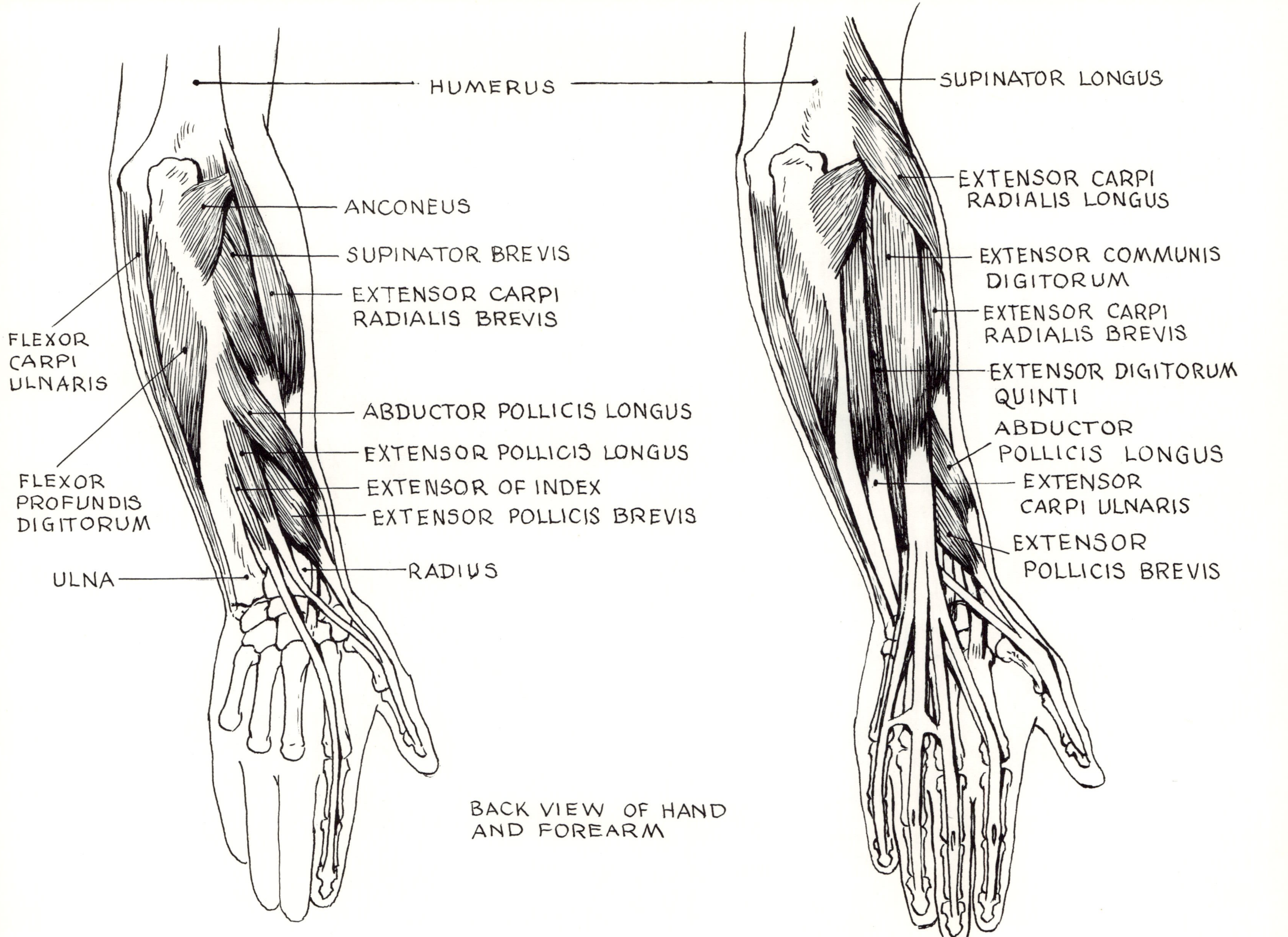

BACK VIEW OF HAND AND FOREARM

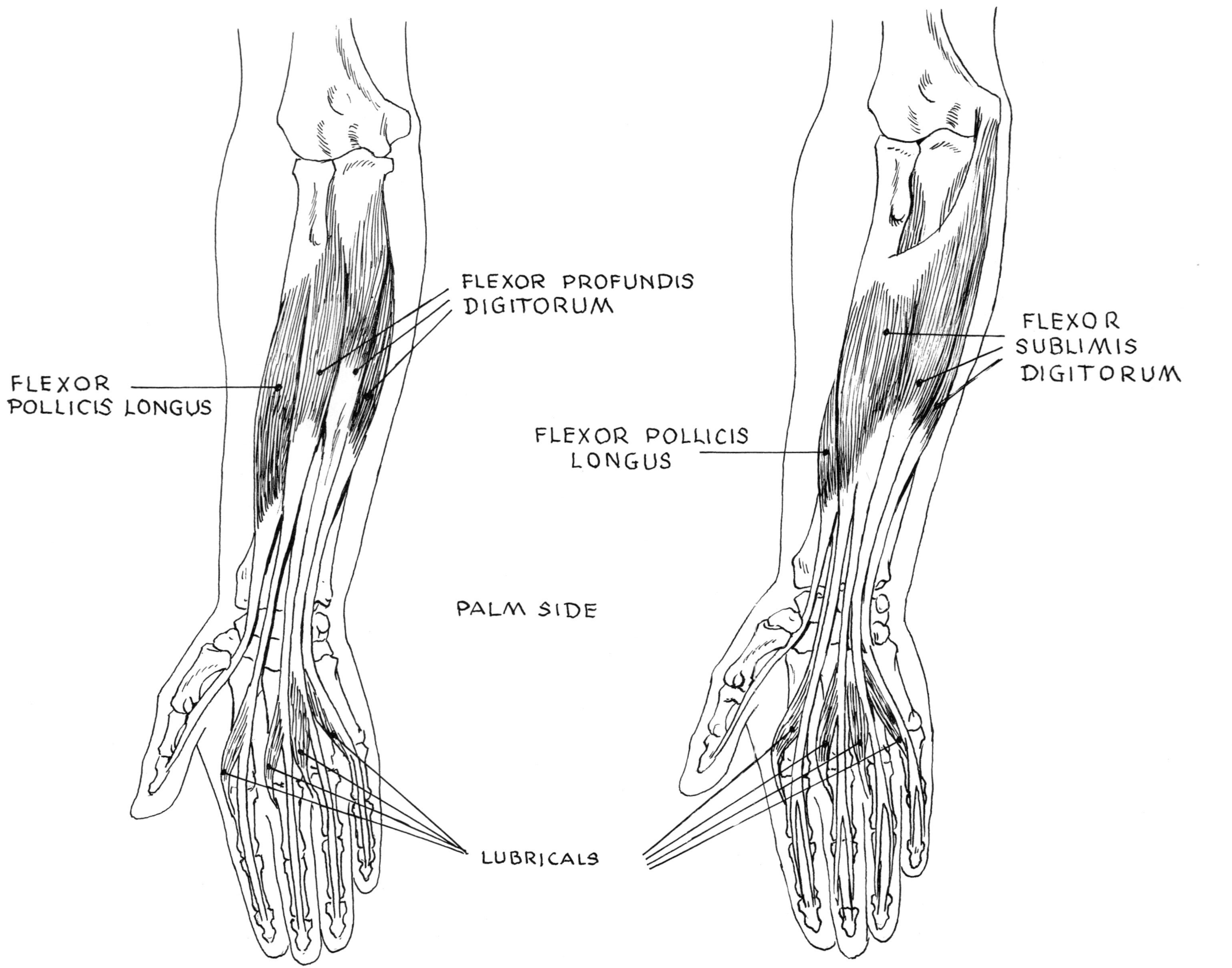
FLEXOR PROFUNDIS
DIGITORUM
FLEXOR
POLLICIS LONGUS
FLEXOR
SUBLIMIS
DIGITORUM
FLEXOR POLLICIS
LONGUS
PALM SIDE
LUBRICALS

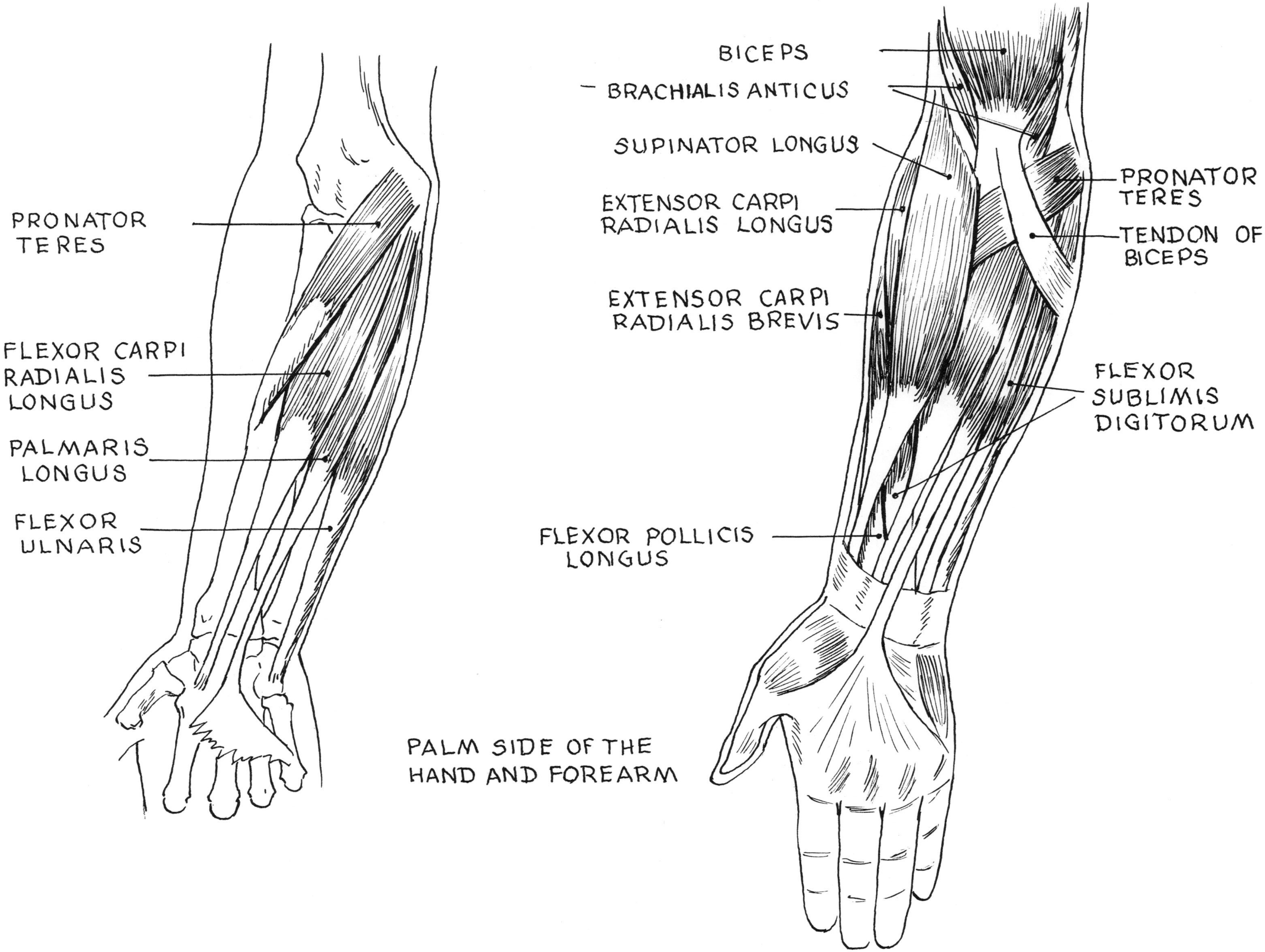

PALM SIDE OF THE HAND AND FOREARM

Thumb side

Little finger side

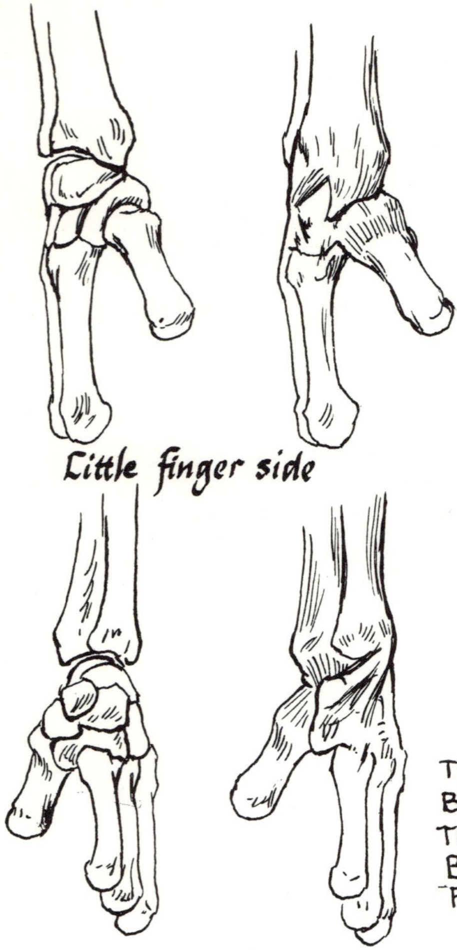

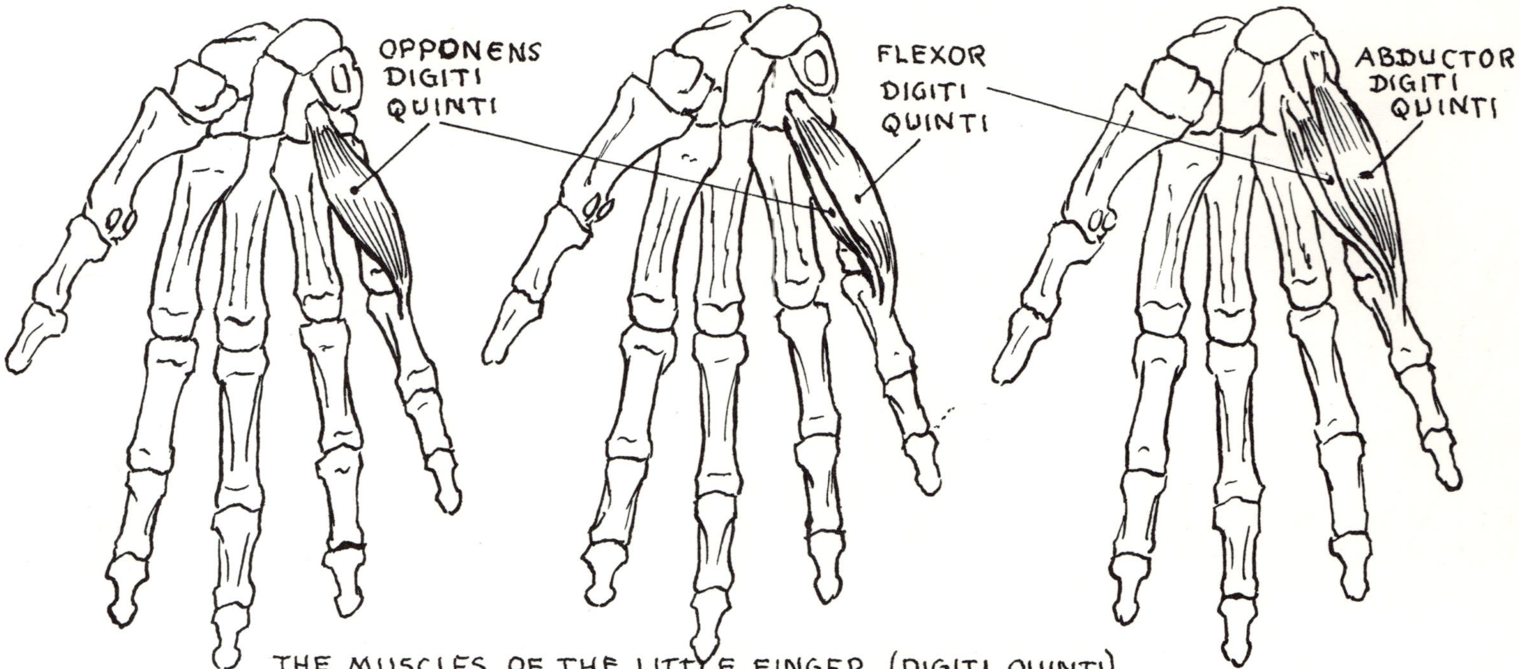

THE MUSCLES OF THE LITTLE FINGER (DIGITI QUINTI) WHICH ORIGINATE IN THE HAND PROPER ARE THE MUSCLES WHICH FORM THE HYPOTHENAR EMINENCE OF THE PALM.

THE TENDONS WHICH BIND THE BONES OF THE WRIST TO THE BONES OF THE FORE ARM

THE INTEROSSEI AND THE LUBRICAL MUSCLES – THE LUBRICALS ARE THE SHADED ONES

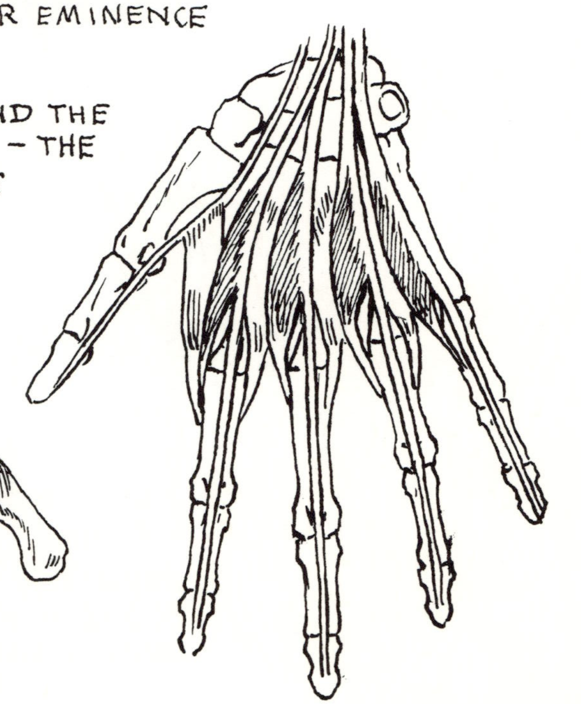

Palm view

Back view

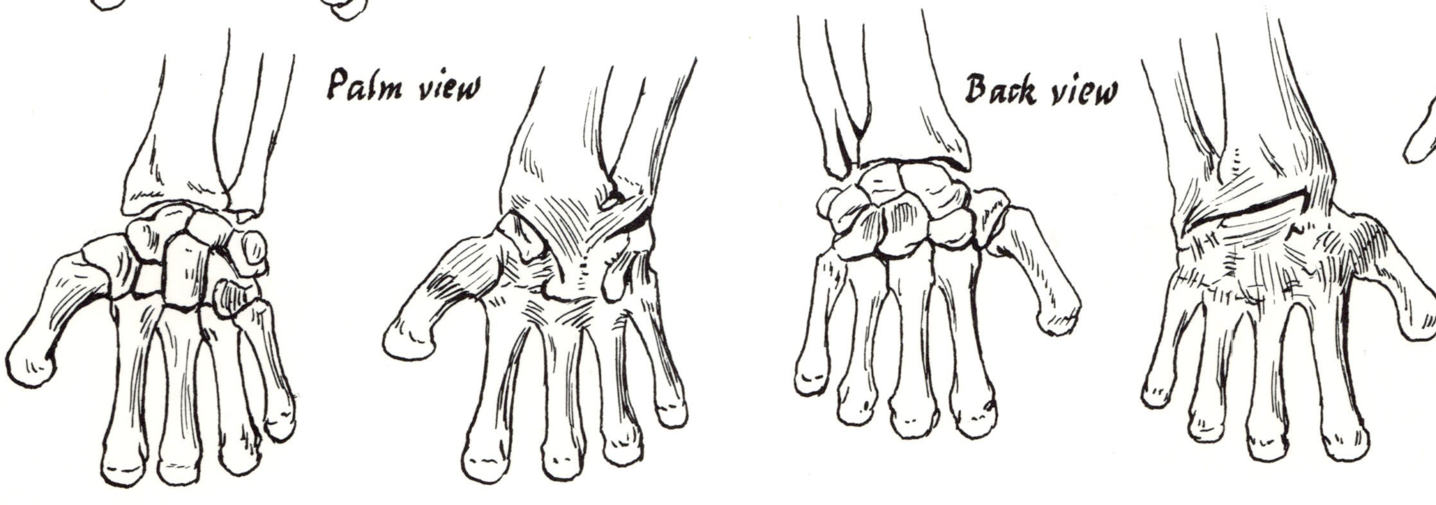

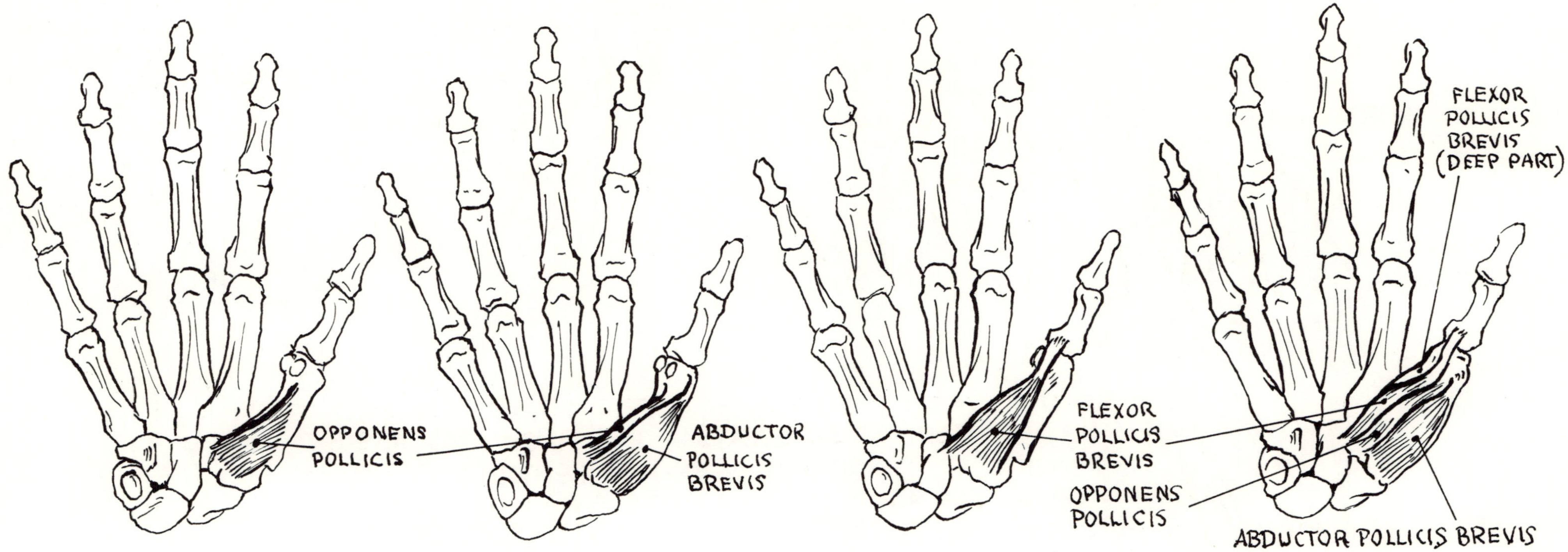

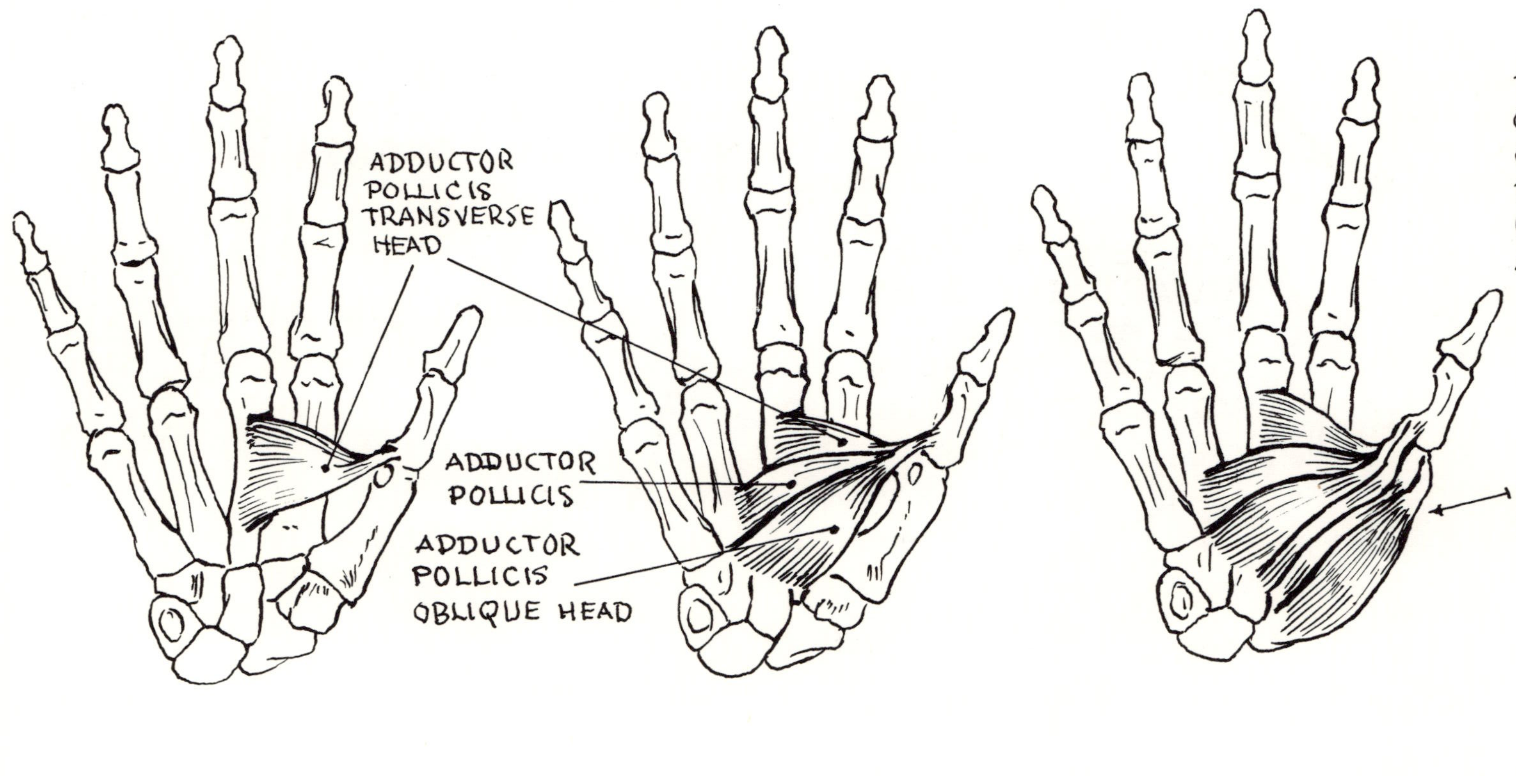

THESE ARE THE MUSCLES OF THE THUMB WHICH ORIGINATE IN THE HAND PROPER. THERE ARE FOUR OTHER MUSCLES THAT ASSIST IN THE CONTROL OF THE THUMB BUT THEY ORIGINATE IN THE FOREARM.

THIS IS THE MASS THAT FORMS THE THENAR EMINENCE OF THE PALM.

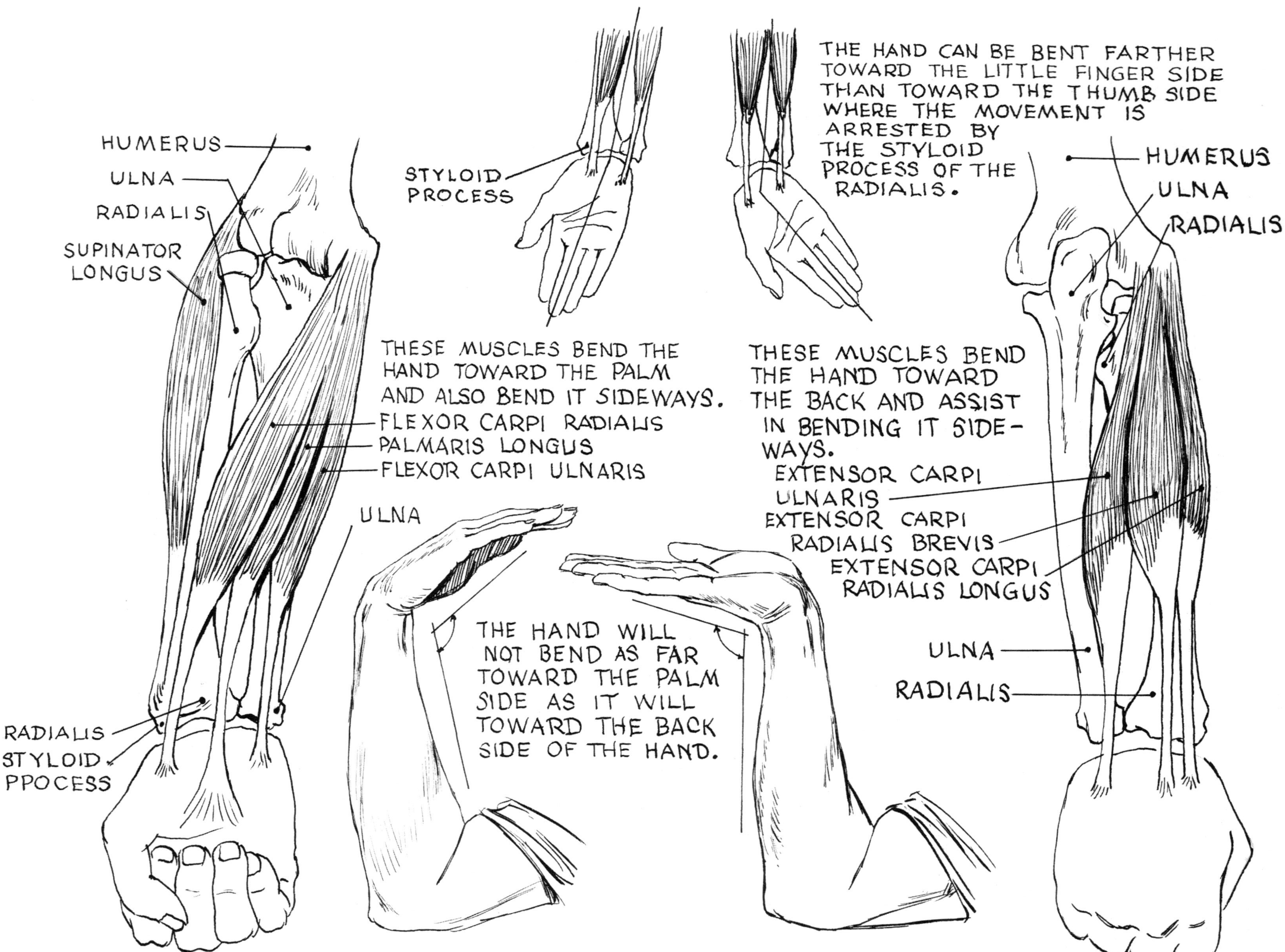
HUMERUS
ULNA
RADIALIS
SUPINATOR LONGUS
RADIALIS
STYLOID PPOCESS
STYLOID PROCESS
THESE MUSCLES BEND THE HAND TOWARD THE PALM AND ALSO BEND IT SIDEWAYS.
FLEXOR CARPI RADIALIS
PALMARIS LONGUS
FLEXOR CARPI ULNARIS
ULNA
THE HAND WILL NOT BEND AS FAR TOWARD THE PALM SIDE AS IT WILL TOWARD THE BACK SIDE OF THE HAND.
THE HAND CAN BE BENT FARTHER TOWARD THE LITTLE FINGER SIDE THAN TOWARD THE THUMB SIDE WHERE THE MOVEMENT IS ARRESTED BY THE STYLOID PROCESS OF THE RADIALIS.
THESE MUSCLES BEND THE HAND TOWARD THE BACK AND ASSIST IN BENDING IT SIDE-WAYS.
EXTENSOR CARPI ULNARIS
EXTENSOR CARPI RADIALIS BREVIS
EXTENSOR CARPI RADIALIS LONGUS
HUMERUS
ULNA
RADIALIS
ULNA
RADIALIS

THE MAJOR TENDONS OF THE HAND

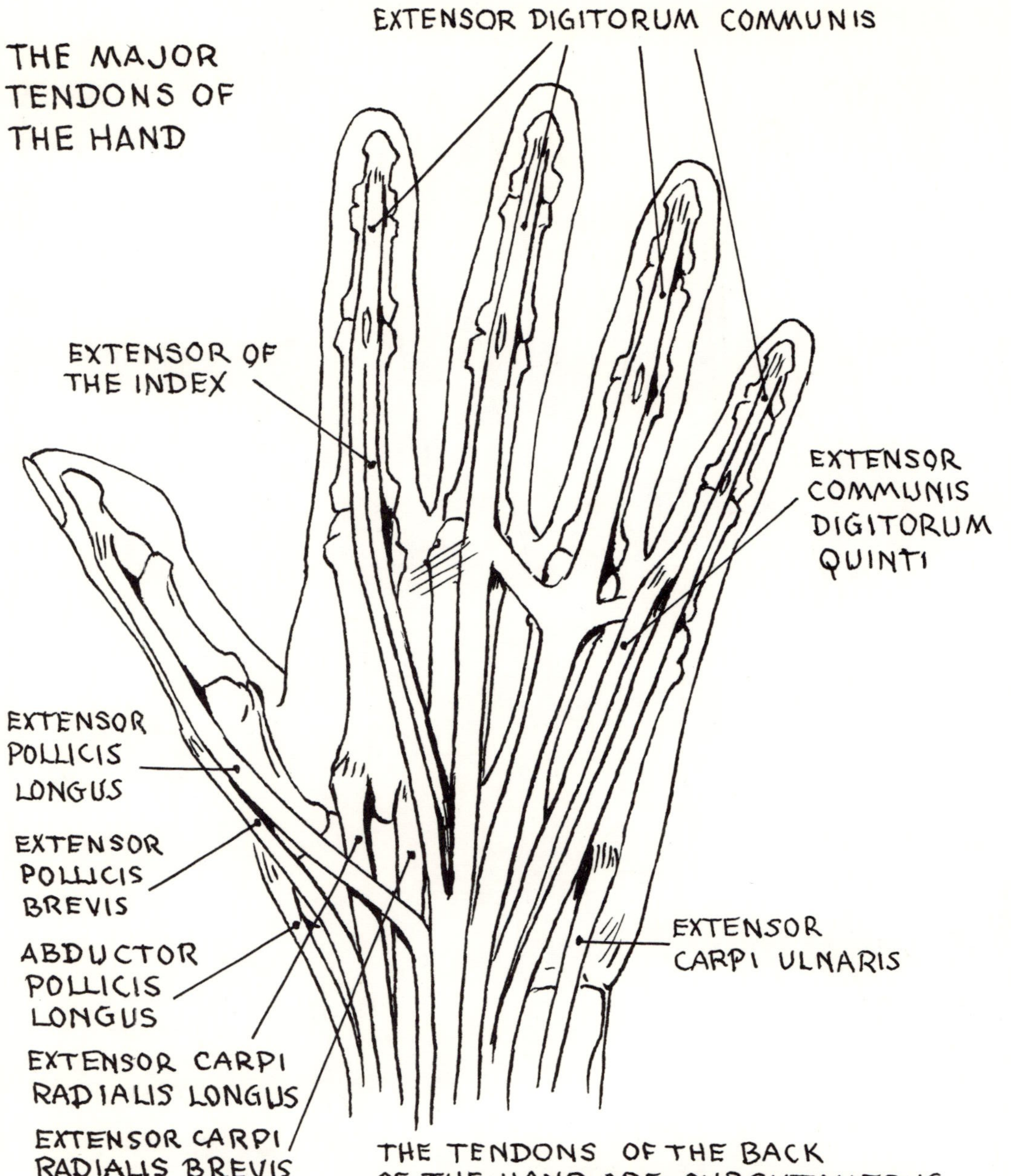

THE TENDONS OF THE BACK OF THE HAND ARE SUBCUTANEOUS (IMMEDIATELY BENEATH THE SURFACE OF THE SKIN) AND CAN BE SEEN CLEARLY WHEN THE FINGERS ARE EXTENDED BUT DISAPPEAR, EXCEPT AT THE KNUCKLES WHEN THE HAND IS CLOSED INTO A FIST.

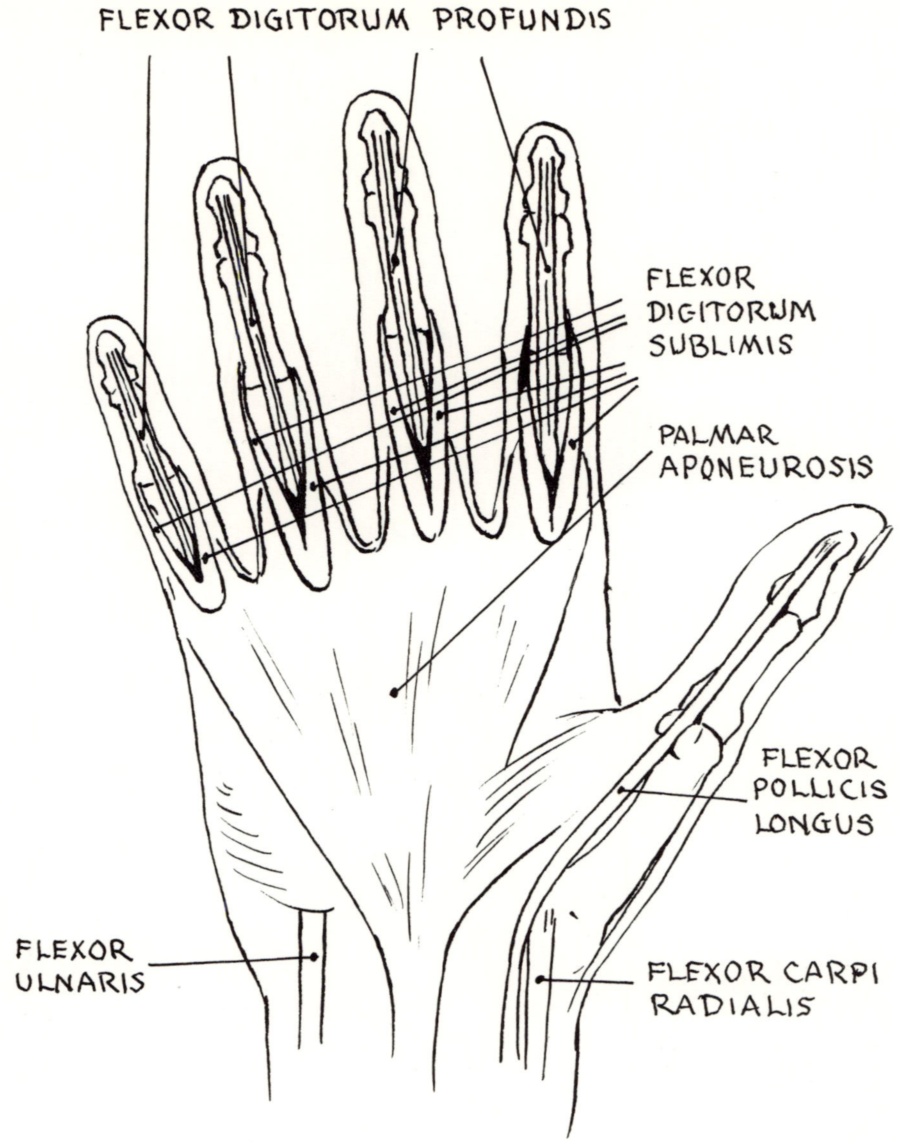

THE TENDONS OF THE PALM SIDE OF THE HAND CAN NOT BE SEEN BECAUSE THEY LIE BENEATH THE HEAVY FLESH PADS OF THE PALM FINGERS AND THUMB.

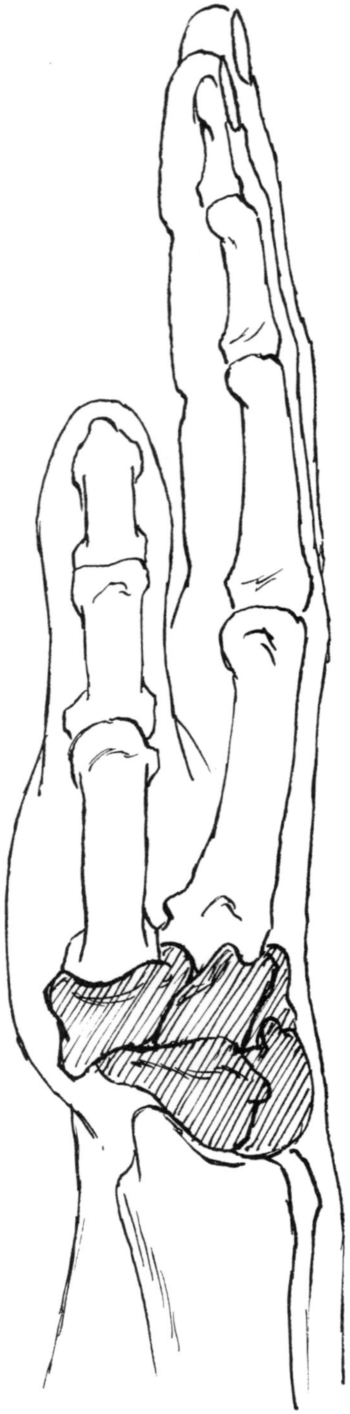

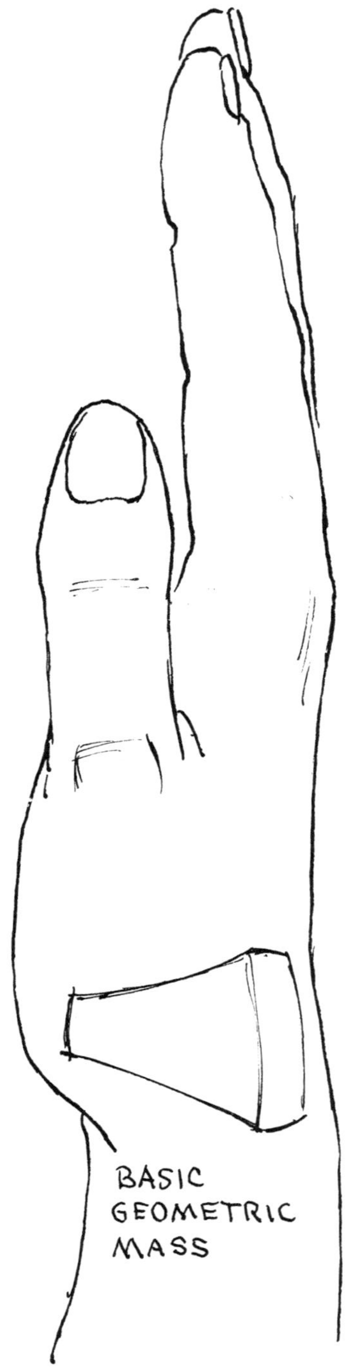

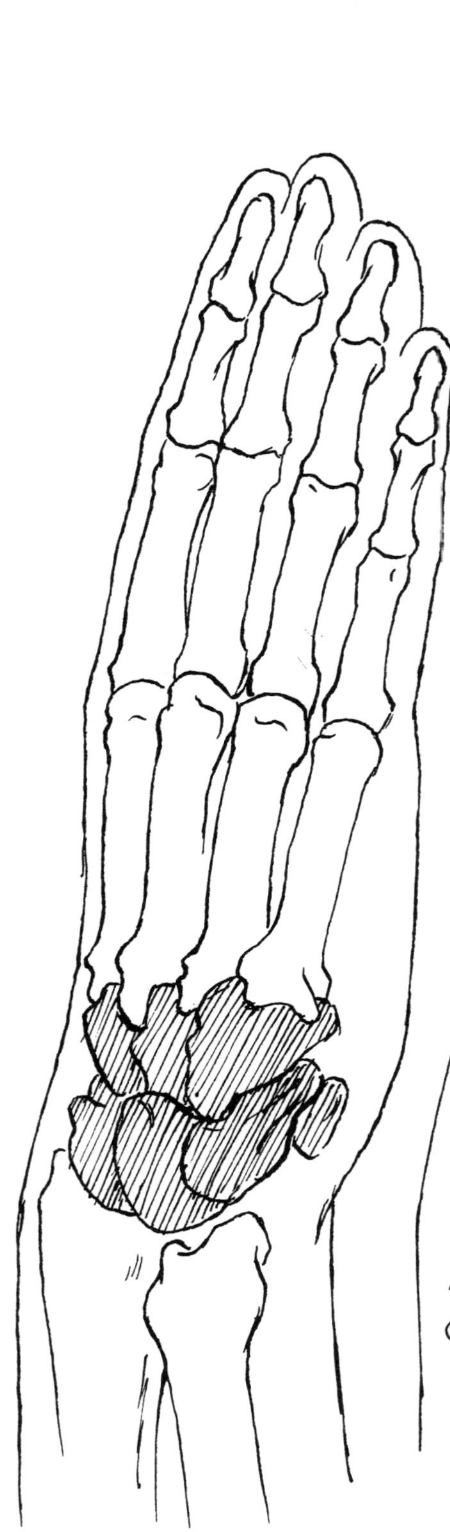

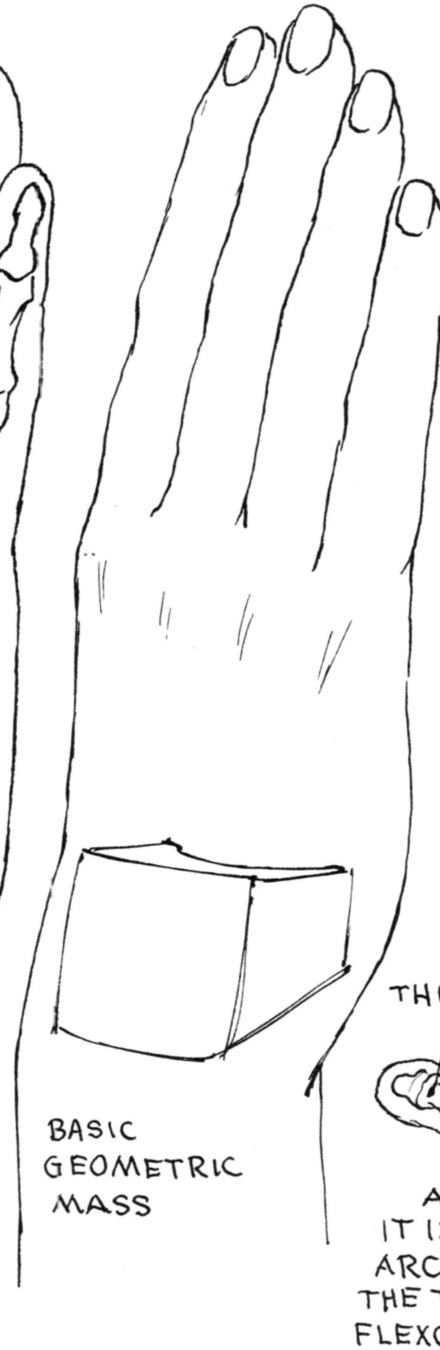

THE EIGHT CARPAL BONES WHICH FORM THE MASS OF THE WRIST MAKE AN ARCHED TRANSITION FROM THE FOREARM INTO THE HAND. THEY ARE NOT EASILY SEEN ON THE SURFACE EXCEPT WHEN THE HAND IS FULLY FLEXED, NAMELY, BENT TOWARD THE PALM. NOTE THE SUBTILE LINE THEY MAKE BETWEEN THE FOREARM AND THE HAND.

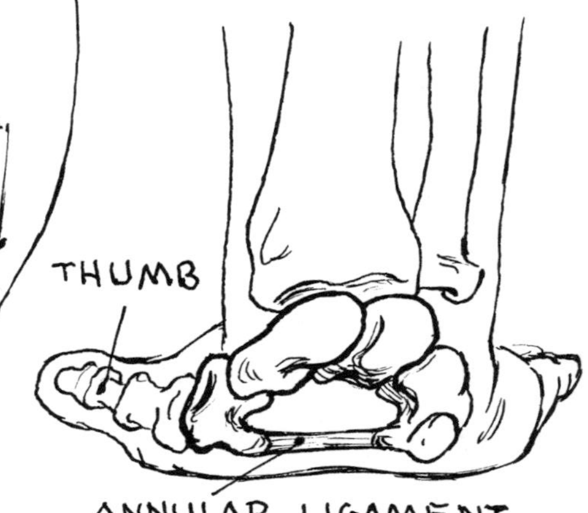

IT IS THROUGH THIS ARCHED CHANNEL THAT THE TENDONS OF THE FLEXOR MUSCLES PASS TO THE PALM AND FINGERS.

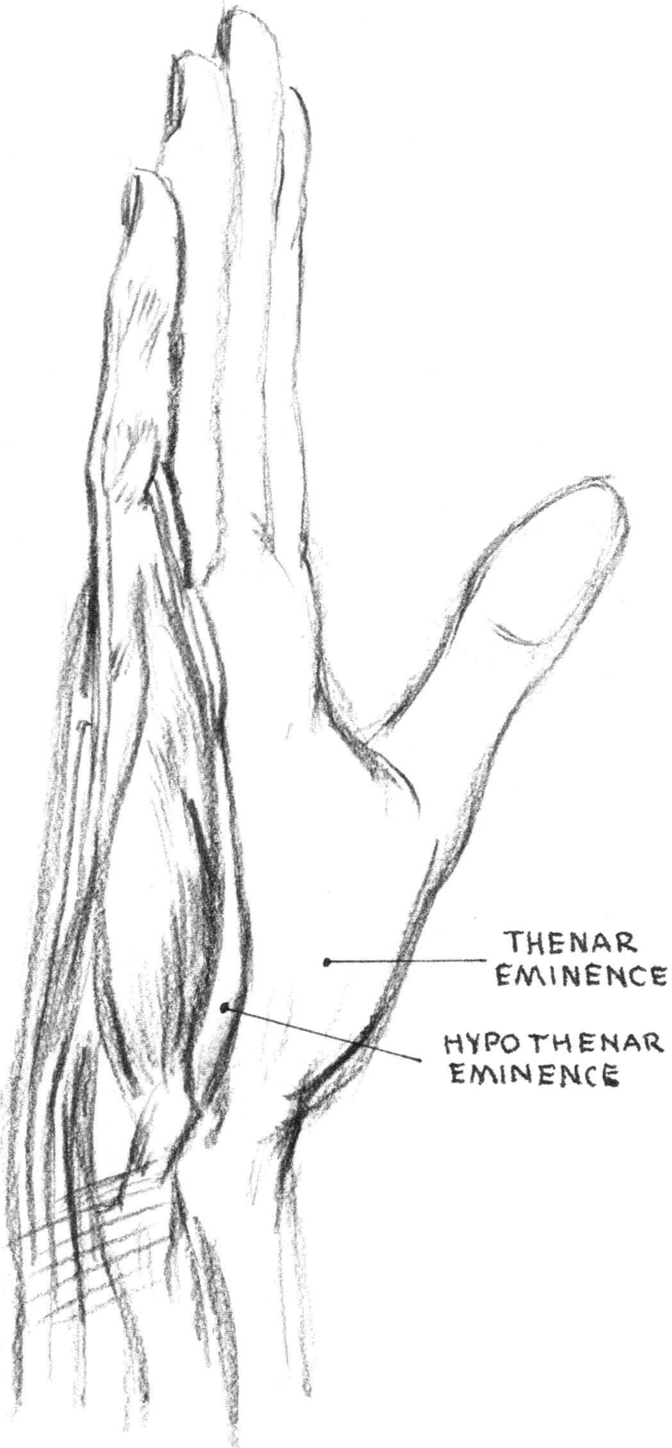

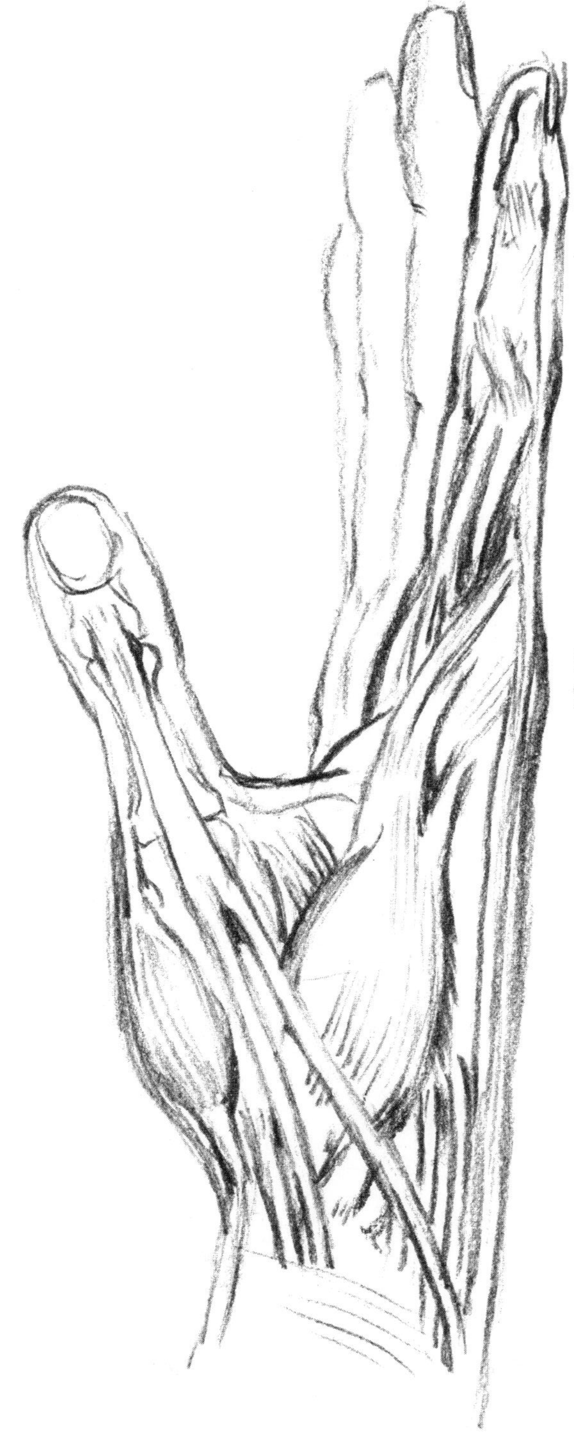

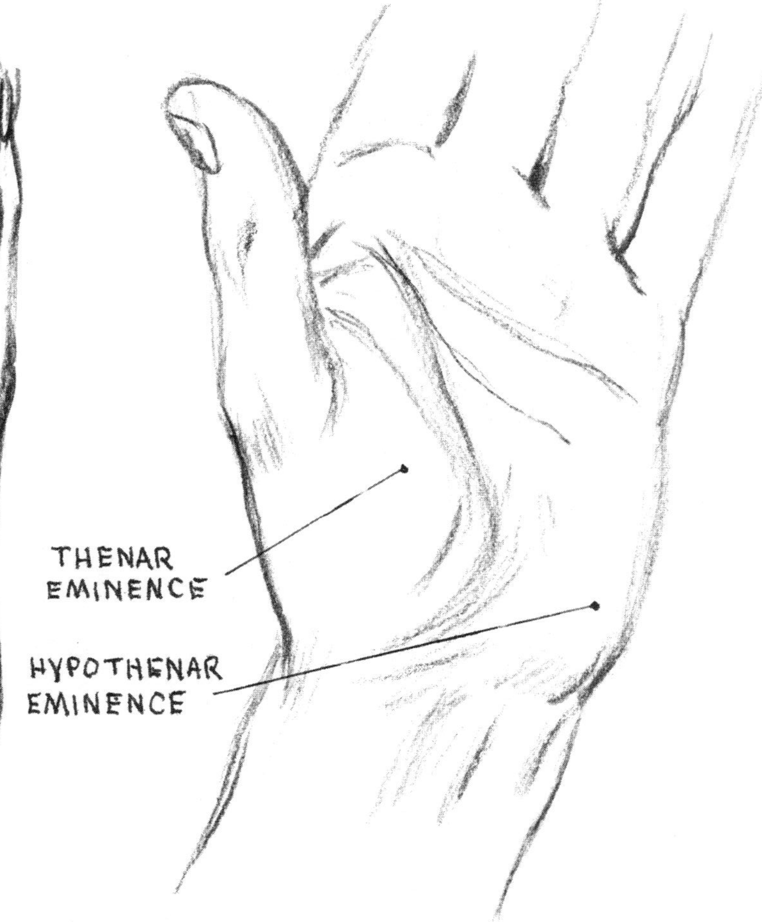

THE ONLY AREAS IN THE HAND WHERE MUSCLES MAKE THEMSELVES EVIDENT ARE BETWEEN THE THUMB AND INDEX FINGER, AT THE BASE OF THE THUMB (THE THENAR EMINENCE) AND BENEATH THE LITTLE FINGER (THE HYPOTHENAR EMINENCE).

THE OTHER MUSCLES THAT ORIGINATE IN THE HAND, NAMELY, THE LUBRICALS AND THE INTEROSSEI ARE HIDDEN BY FLESHY PADS AND DO NOT SHOW ON THE SURFACE.

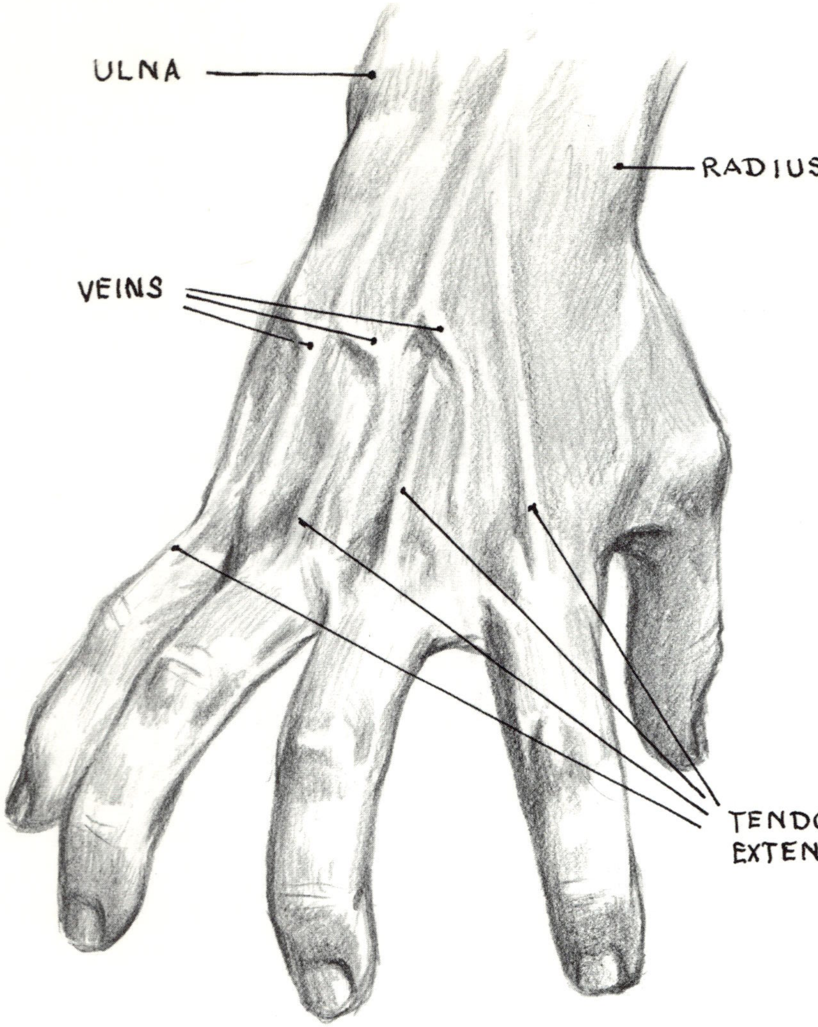

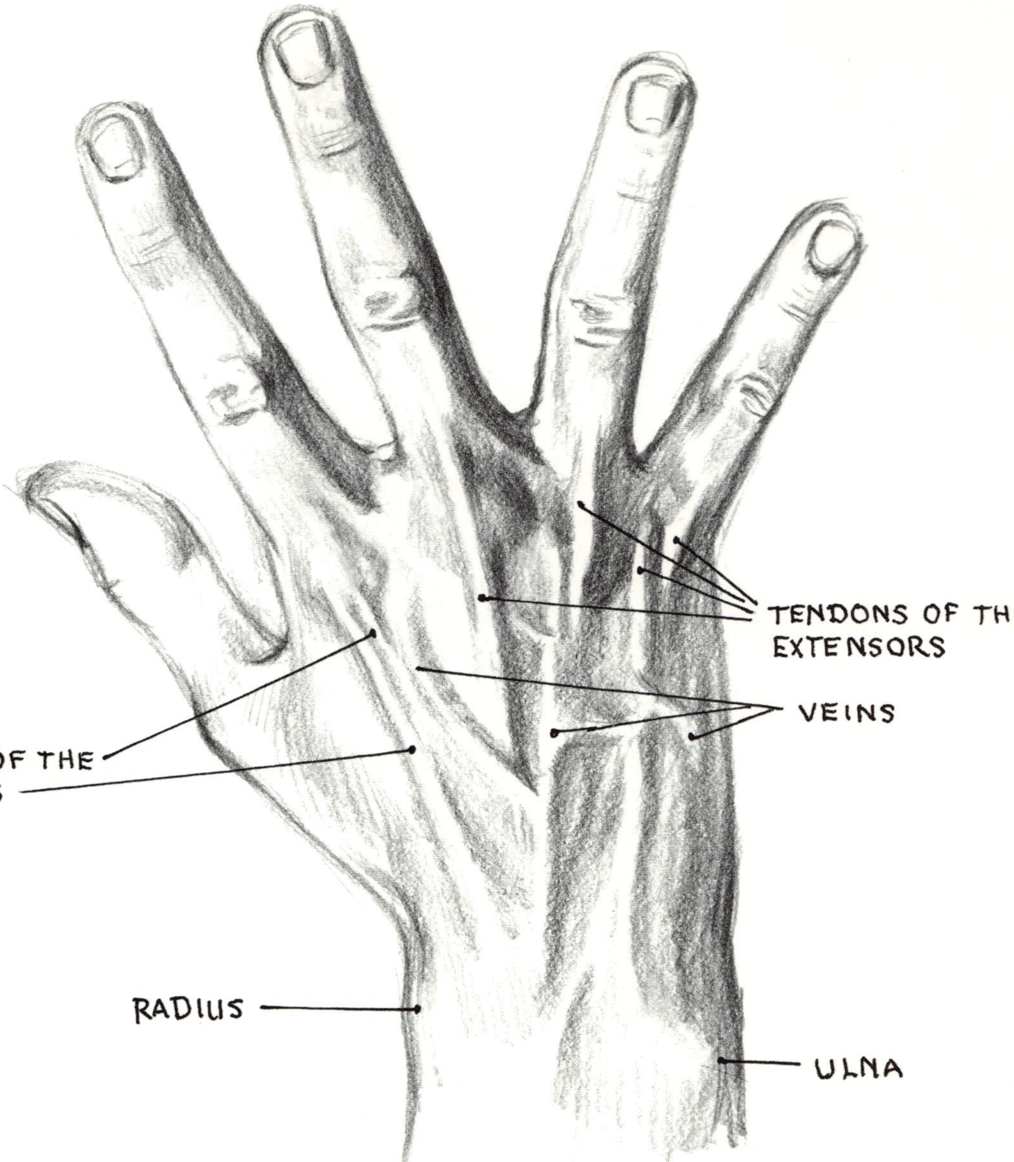

THIS DRAWING SHOWS A DUAL ACTION, NAMELY, AN EXTENSION OF THE FINGERS AND A FLEXION OF THEIR TIPS. THE VEINS WILL VARY GREATLY WITH DIFFERENT HANDS BEING HARDLY DISCERNIBLE IN A FAT HAND.

HERE THE FINGERS ARE FULLY EXTENDED AND THE TENDONS OF THE EXTENSOR MUSCLES ARE CLEARLY SEEN.

DRAWN ON HEAVY BOND PAPER WITH 2B GRAPHITE PENCIL.

THE LATIN NAMES OF THE MUSCLES OF THE FOREARM AND THE HAND WITH THE ENGLISH TRANSLATION AND THEIR ACTION

Latin	English	Action
Flexor carpi radialis	Bender of the wrist on the radius bone side (thumb side)	Bends the wrist toward the palm side
Flexor carpi ulnaris	Bender of the wrist on the ulna bone side (little finger side)	Bends the wrist toward the palm side
Flexor digitorum profundis	Bender of the fingers Deep muscle	Bends the tips of the fingers
Flexor digitorum sublimis	Bender of the fingers muscle beneath the surface of the skin	Bends the second joint of the fingers
Flexor longus pollicis	Long bender of the thumb	Bends the end joint of the thumb
Palmaris longus	Long palm muscle	Helps bend the wrist
Pronator teres	Round turner or rotator	Rotates the hand palm downward
Pronator guadratus	Square turner or rotator	Helps rotate the hand palm down
Supinator longus	The long muscle that lays the palm up-ward	Rotates the hand so that the palm is turned up and helps bend the elbow
Extensor carpi radialis longus	The long straightener of the wrist on the radius bone side	Straightens the wrist and assists in bend-ing it toward the thumb side

Latin	English	Action
Extensor carpi radialis brevis	The short straightener of the wrist on the radius bone side (thumb side)	Straightens the wrist and assists in bending it sidewards toward the thumb side
Extensor communis digitorum	The straightener that communicates with the fingers	Straightens the fingers
Extensor digiti quinti	Little finger straightener	Straightens the little finger independently
Extensor carpi ulnaris	Straightener of the wrist on the ulna bone side (little finger side)	Straightens the wrist and also bends it sideways on the little finger side
Anconeus	Elbow muscle	Helps straighten the forearm
Supinator brevis	Short muscle that helps turn the palm face upward	Helps the supinator longus turn the palm face up
Extensor pollicis brevis	Short straightener of the thumb	Straightens the thumb
Extensor pollicis longus	Long straightener of the thumb	Straightens the thumb
Extensor index	Straightener of the index finger	Straightens the index finger independently

THE LATIN AND GREEK NAMES OF THE BONES OF THE FOREARM AND HAND WITH THEIR ENGLISH TRANSLATION

Ulna (Latin)	Latin word for elbow. The bone in the forearm that forms the elbow and extends to the wrist on the little finger side.	Metacarpals (Latin)	Beyond the wrist. Four of these bones form the bonal structure of the back and palm, the fifth metacarpal forms the base of the thumb.
Radius (Latin)	Latin word for spoke of a wheel. The bone that starts at the side of the ulna at the elbow and extends to the wrist and turns across the Ulna when the hand is rotated	Phalanges (Latin)	From the Latin word phalanx, a body of soldiers lined one behind the other. There are fourteen phalanges, two in the thumb and three in each of the fingers.
Carpals (Greek) The following eight words are the names of the individual bones of the wrist.	From the Greek word *karpos* meaning wrist.		
Trapezium (Greek)	An irregular four-sided shape.		
Trapezoid (Greek)	A plane four-sided figure with two parallel sides.		
Scaphoid (Greek)	Boat shaped		
Os magnum (Latin)	Large bone		
Semi-lunar (Latin)	Half-moon shaped		
Unciform (Latin)	Hook shaped		
Cuneiform (Latin)	Wedge shaped		
Pisiform (Latin)	Pea shaped		

Latin	English	Action
Abductor pollicis brevis	Short puller away of the thumb	Pulls the thumb away from the hand
Opponens pollicis	Puller of the thumb	Pulls the thumb across the palm
Flexor pollicis brevis	Short bender of the thumb	Bends the thumb
Adductor pollicis longus and oblique head	Puller of the thumb toward the palm — long and slanting head	Pulls the thumb across the palm
Palmaris brevis	Short palmar muscle	Contracts palmar skin
Abductor digiti quinti	Puller away of the little finger	Pulls the little finger to the side away from the other fingers
Flexor brevis digiti quinti	Short bender of the little finger	Bends the little finger independently
Opponens digiti quinti	Puller of the little finger.	Draws the little finger toward the other fingers
Lumbricals (four)	Proper Latin name untranslatable	Bends the first phalanges of the fingers
Dorsal interossei (four)	Back between the bones muscles	Pulls the fingers away from one another
Palmar interossei (three)	Palm between the bones muscles	Closes the fingers together

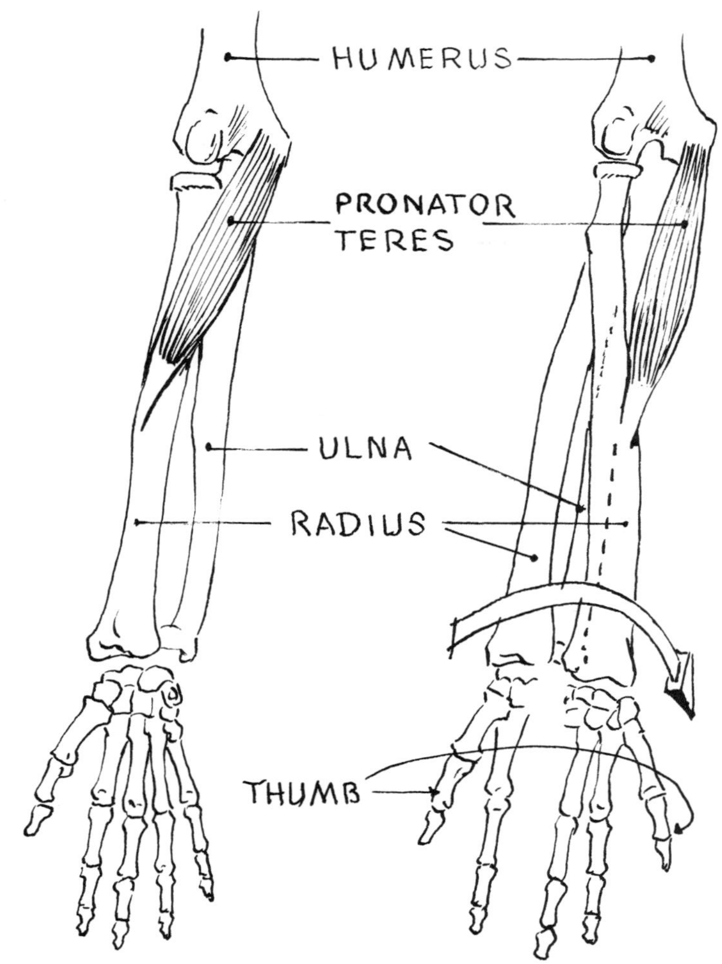

THE MUSCLE WHICH ROTATES THE RADIUS BONE AROUND THE ULNA AND PERMITS US TO TURN OUR HAND PALM DOWN IS THE PRONATOR TERES. THE ULNA DOES NOT MOVE IN THIS ACTION.

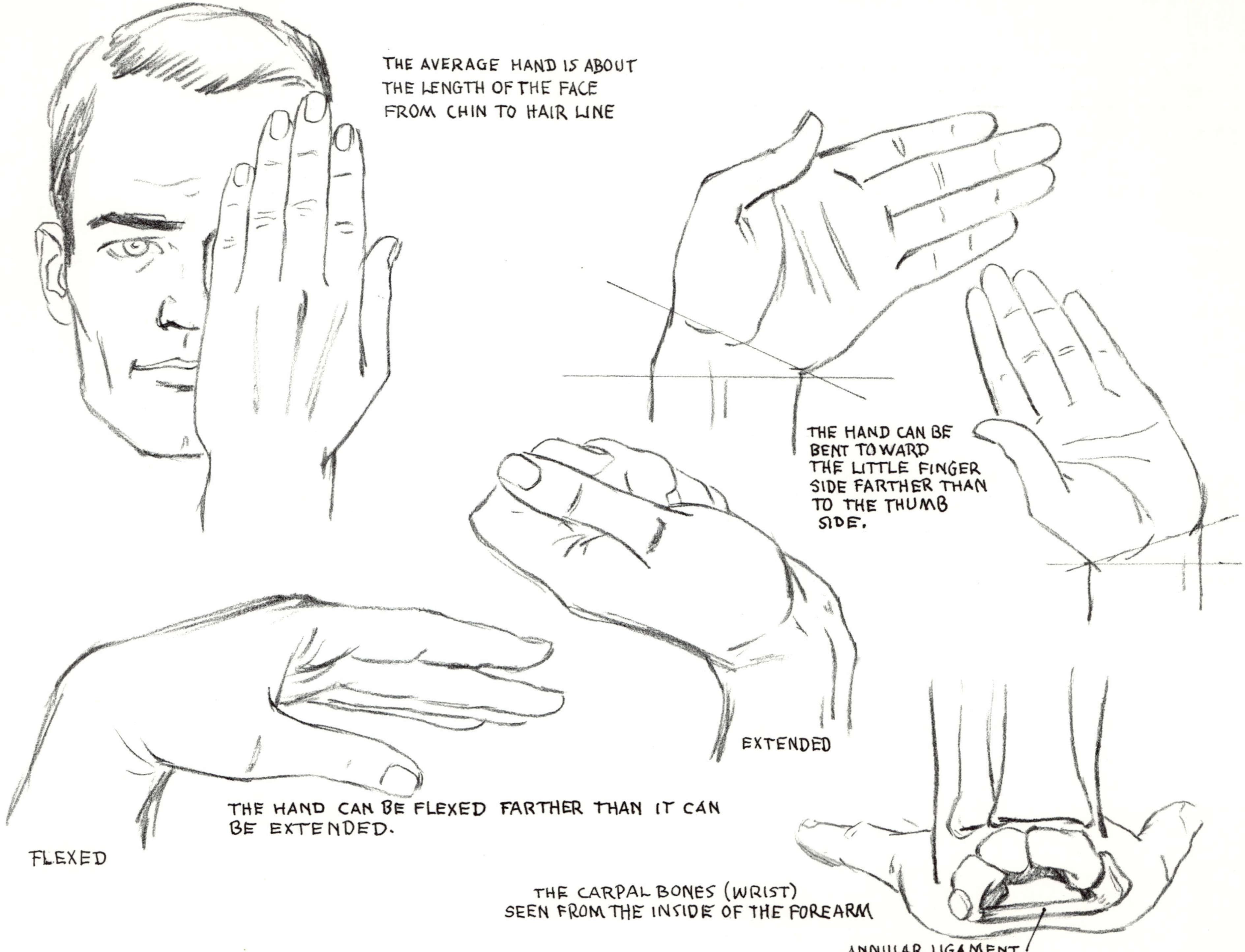

DRAWN ON LIGHT WEIGHT BOND WITH HB WOLFF PENCIL.

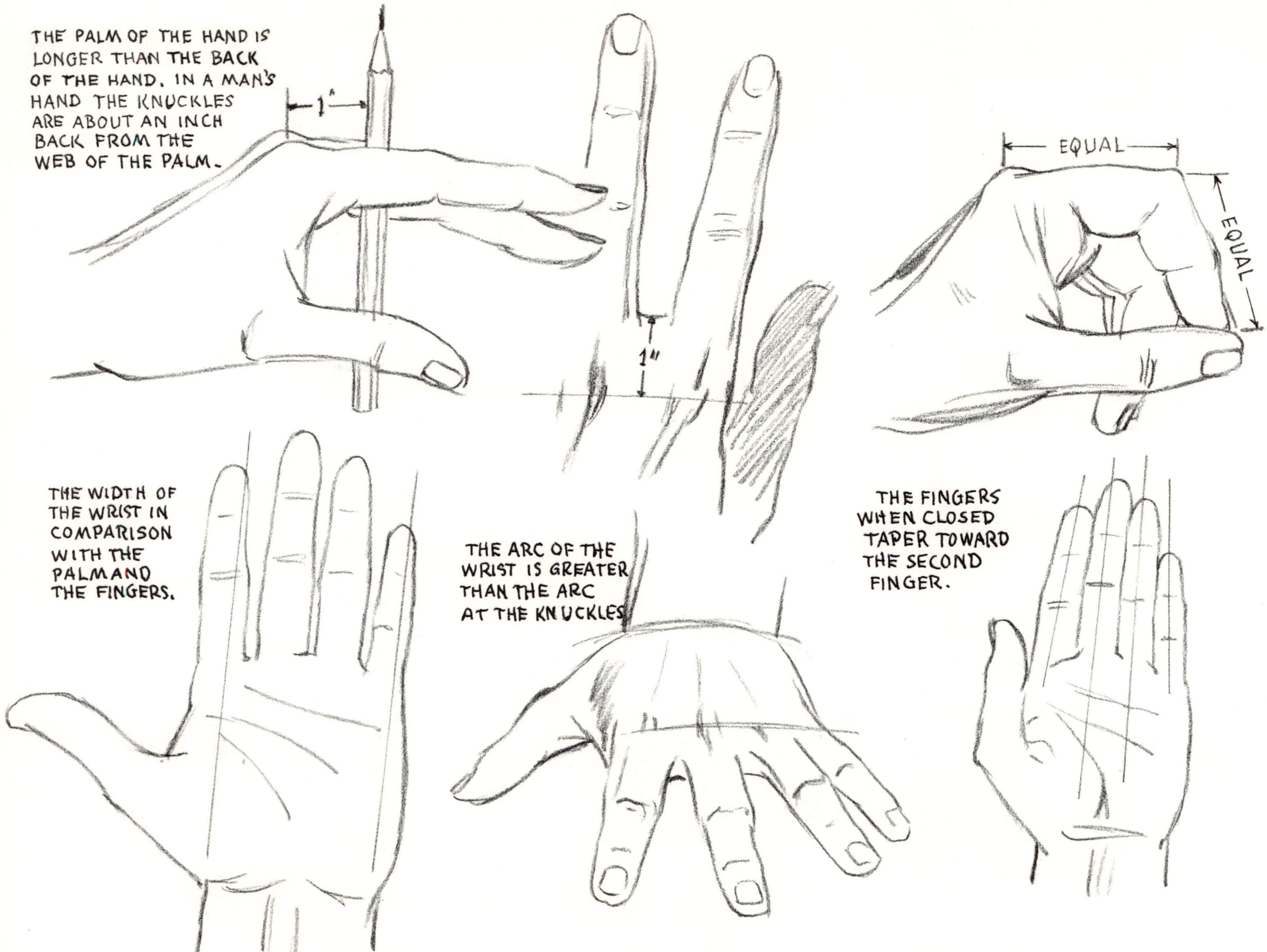

DRAWN ON LIGHT WEIGHT BOND PAPER WITH HB WOLFF PENCIL.

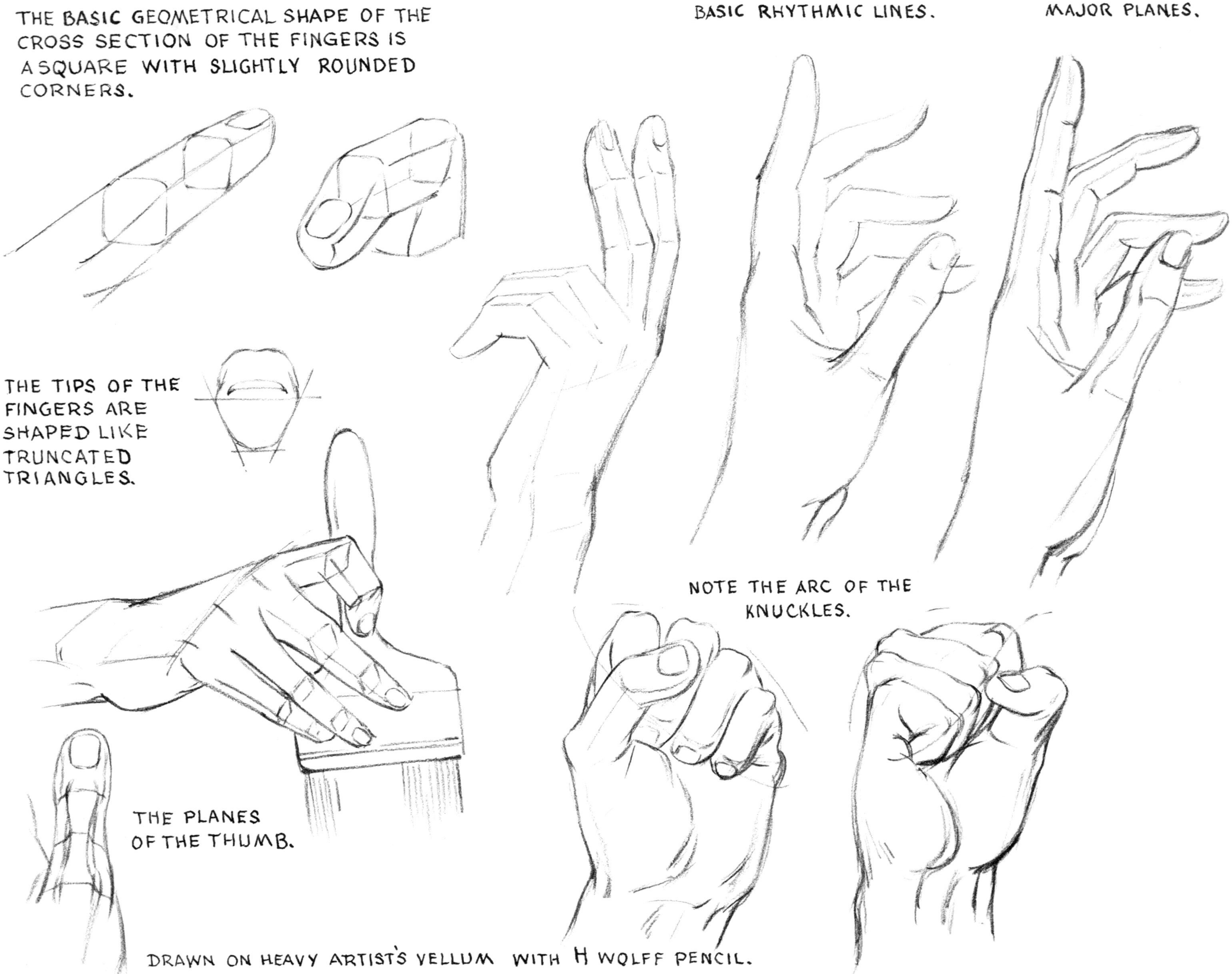

DRAWN ON HEAVY ARTIST'S VELLUM WITH H WOLFF PENCIL.

IF THE BASIC SHAPES OF THE WRIST AND HAND WERE TO BE REDUCED TO BLOCK-LIKE UNITS THEY WOULD ASSUME SHAPES SIMILAR TO THESE DRAWINGS.

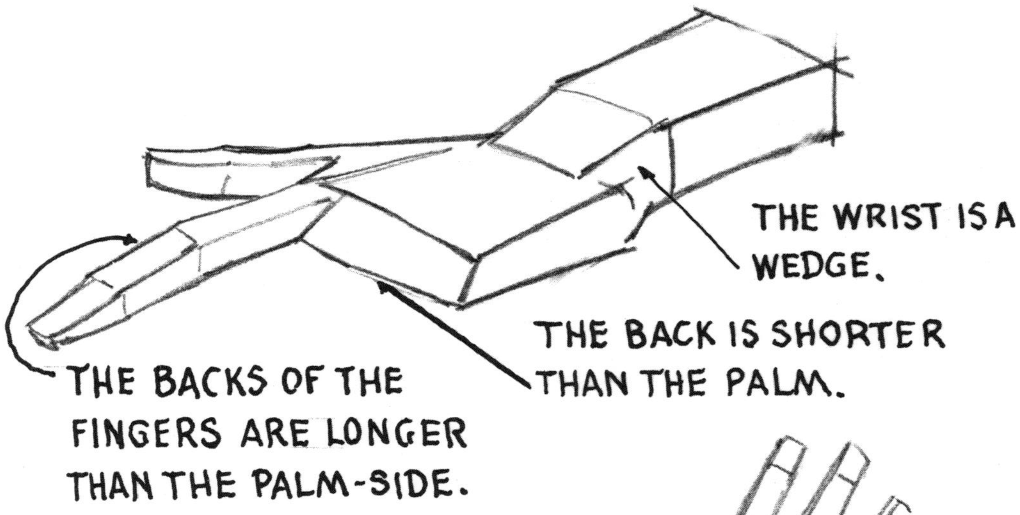

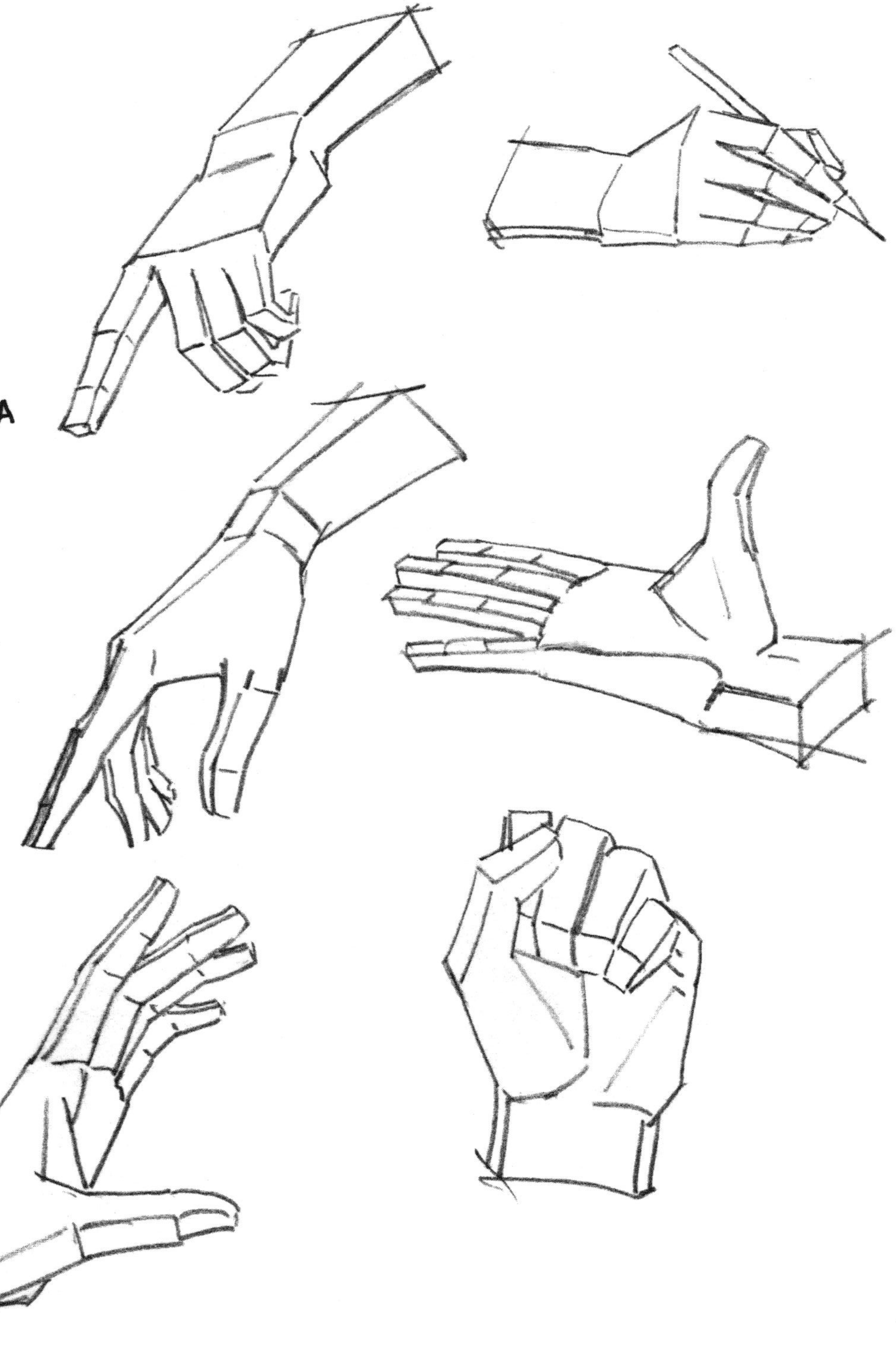

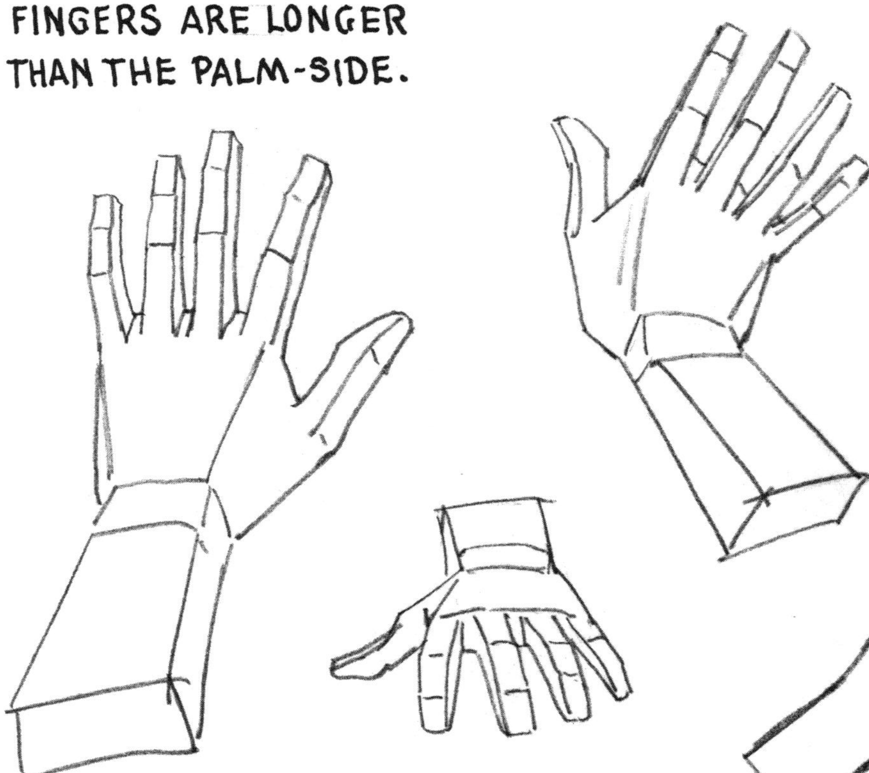

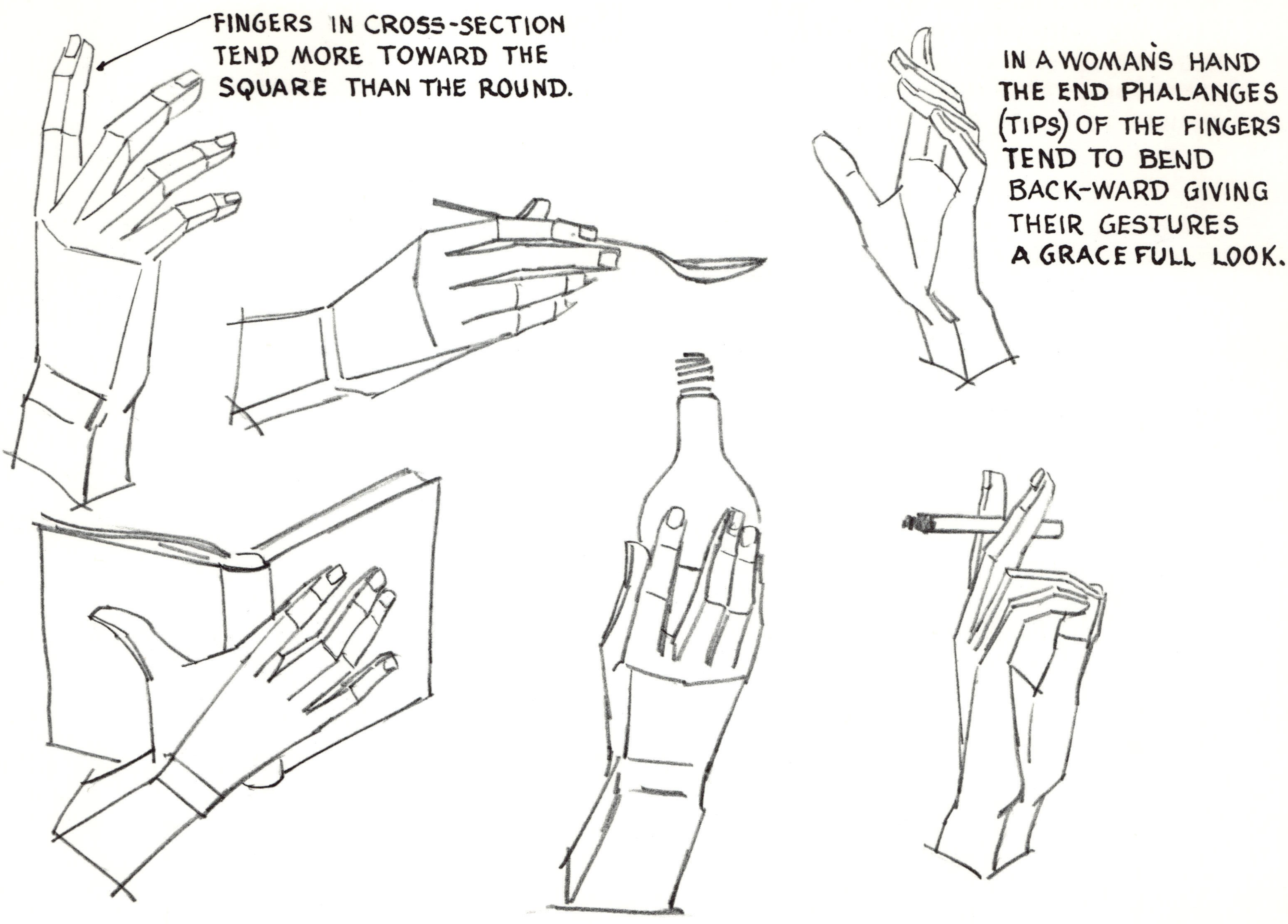
FINGERS IN CROSS-SECTION TEND MORE TOWARD THE SQUARE THAN THE ROUND.
IN A WOMAN'S HAND THE END PHALANGES (TIPS) OF THE FINGERS TEND TO BEND BACK-WARD GIVING THEIR GESTURES A GRACE FULL LOOK.

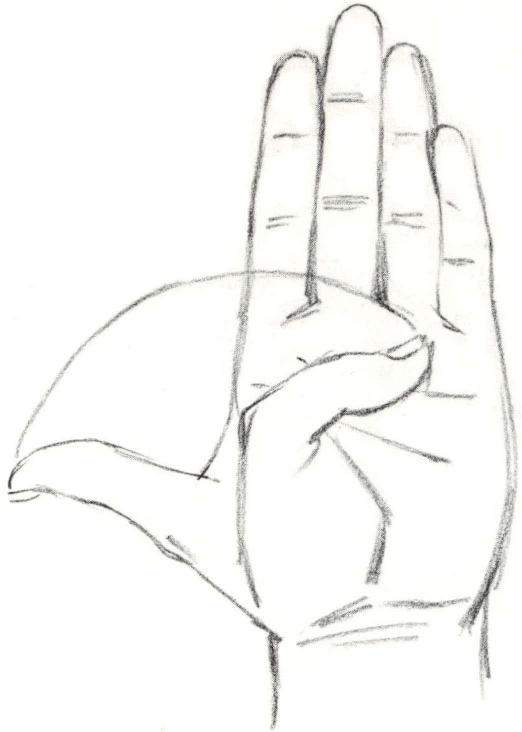

THE THUMB IS THE "POLICEMAN" OF THE HAND

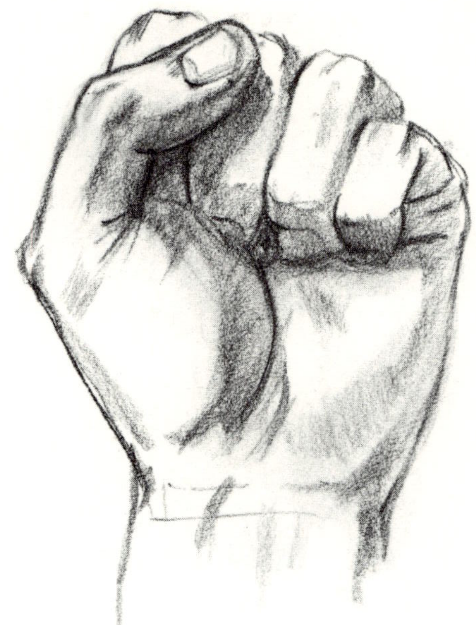

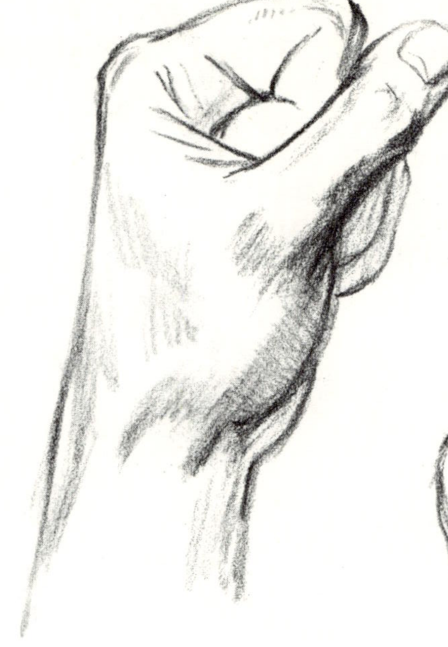

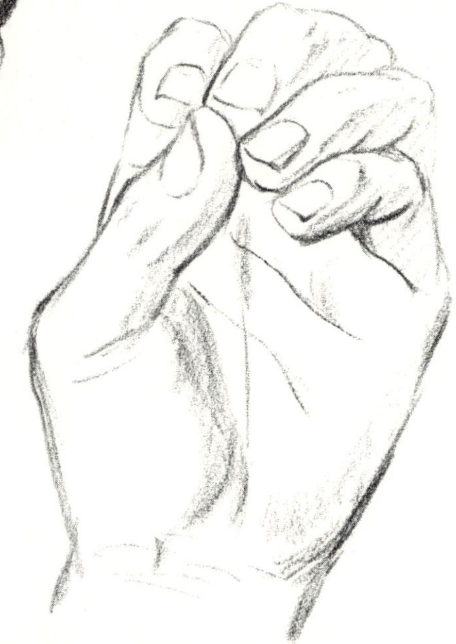

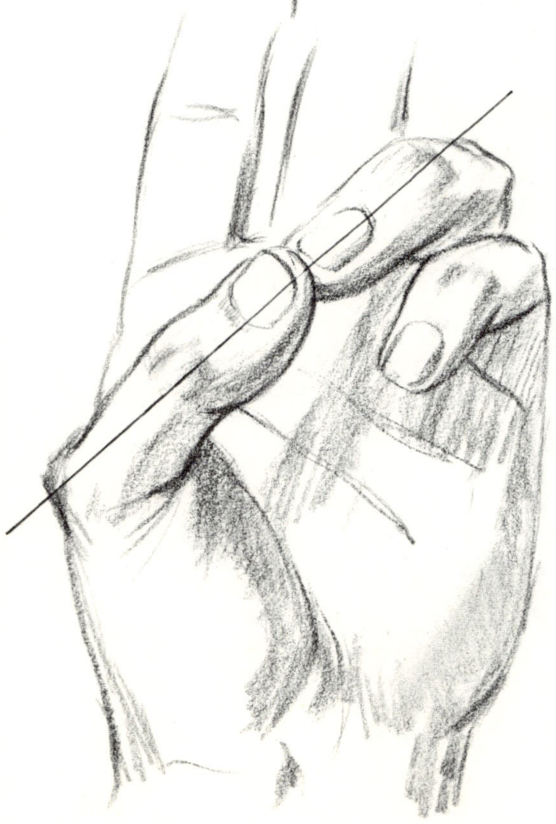

WHEN THE THIRD FINGER AND THE THUMB ARE OPPOSED THEY FORM A STRAIGHT LINE.

ONLY MAN CAN BRING HIS FINGERS AND THUMB TOGETHER. THIS ABILITY HAS MADE MAN THE MOST DEXTEROUS OF ALL ANIMALS.

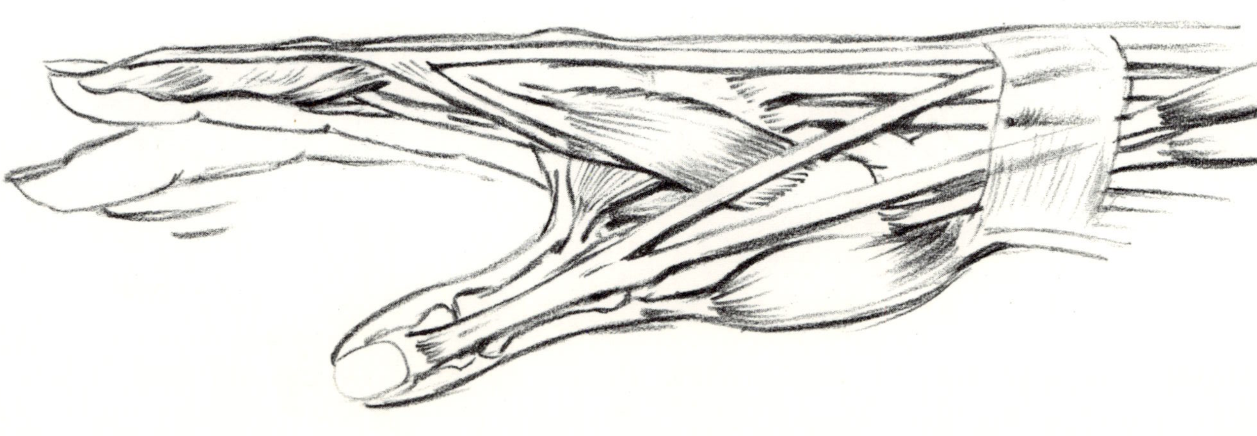

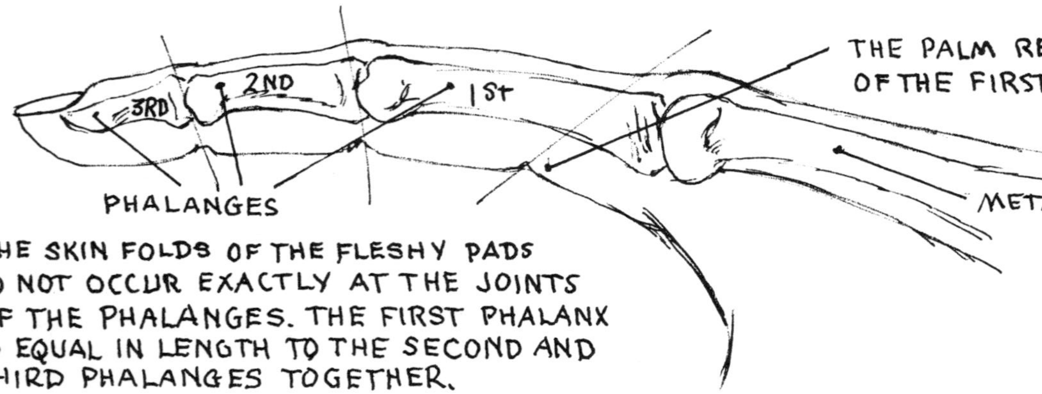

THE PALM REACHES ALMOST TO THE MIDPOINT OF THE FIRST PHALANX OF EACH FINGER.

THE SKIN FOLDS OF THE FLESHY PADS DO NOT OCCUR EXACTLY AT THE JOINTS OF THE PHALANGES. THE FIRST PHALANX IS EQUAL IN LENGTH TO THE SECOND AND THIRD PHALANGES TOGETHER.

WHERE THE INDEX FINGER JOINS THE PALM THERE IS ONLY ONE SKIN FOLD LINE WHILE THE REMAINING THREE FINGERS HAVE DOUBLE LINES. THE LINES IN THE PALM VARY WITH EACH INDIVIDUAL BUT CLOSELY FOLLOW THE PATTERN SHOWN BELOW.

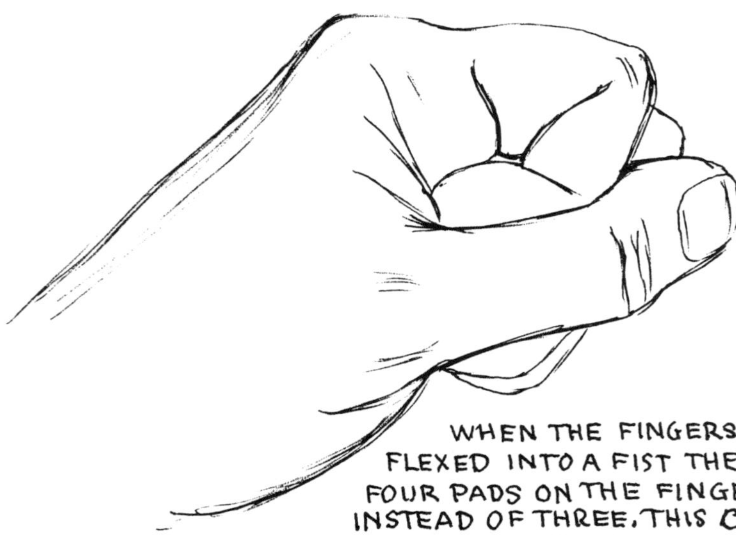

WHEN THE FINGERS ARE FLEXED INTO A FIST THERE ARE FOUR PADS ON THE FINGERS INSTEAD OF THREE. THIS CAN BE SEEN ON THE INDEX AND LITTLE FINGER SIDES OF THE HAND.

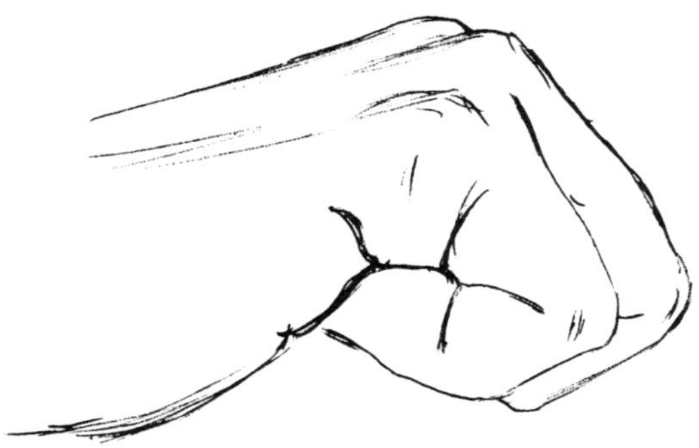

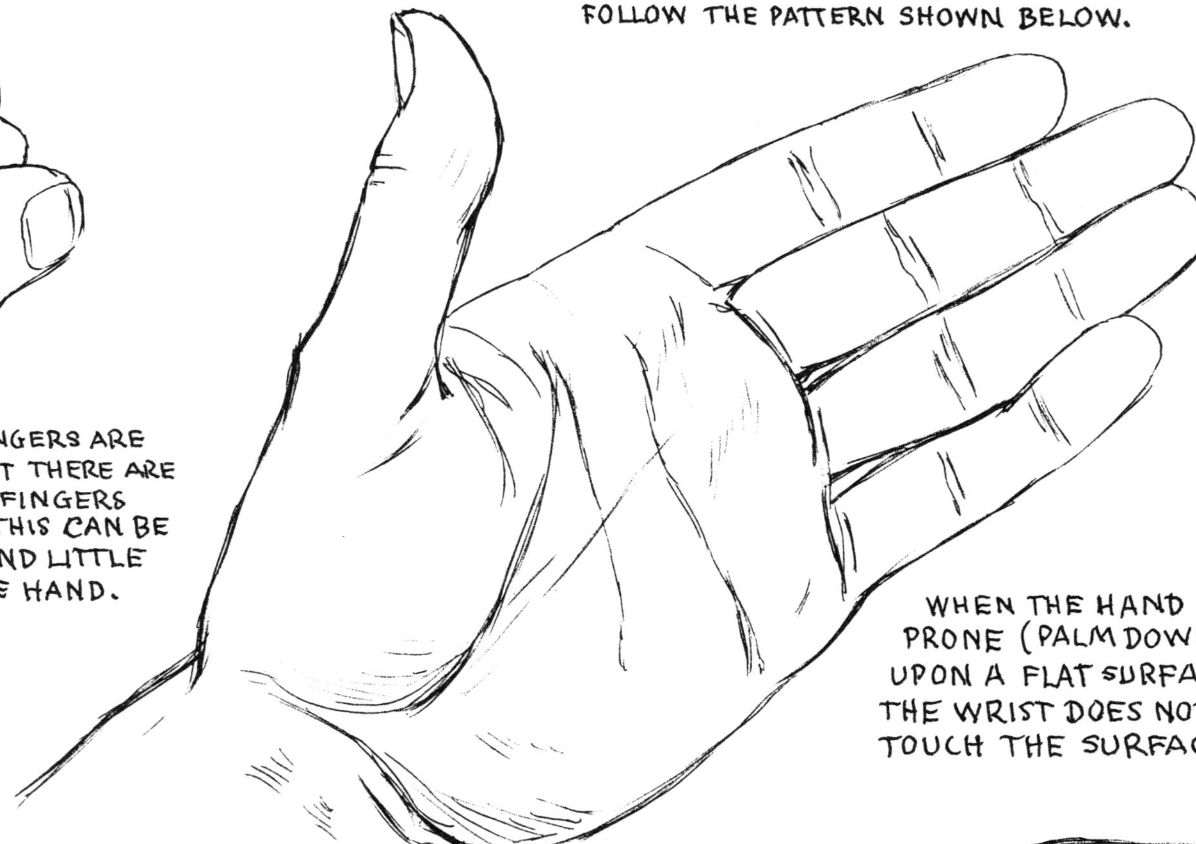

WHEN THE HAND IS PRONE (PALM DOWN) UPON A FLAT SURFACE THE WRIST DOES NOT TOUCH THE SURFACE.

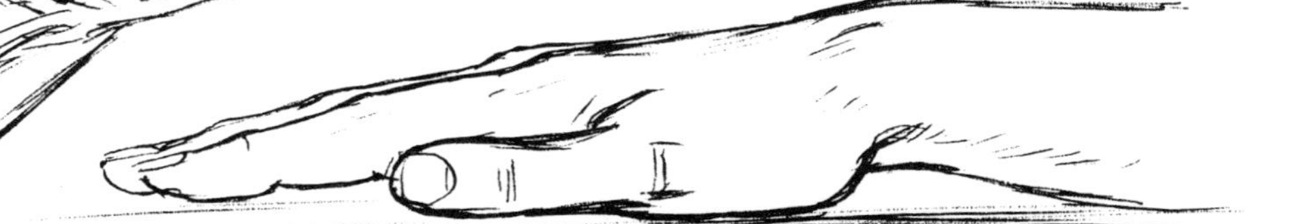

DRAWN ON MEDIUM WEIGHT BOND PAPER WITH A COMMON BALLPOINT PEN.

QUICK SKETCHES (1 MINUTE) TO ESTABLISH A GESTURE
THIS IS A GOOD PRACTICE TO DEVELOP.

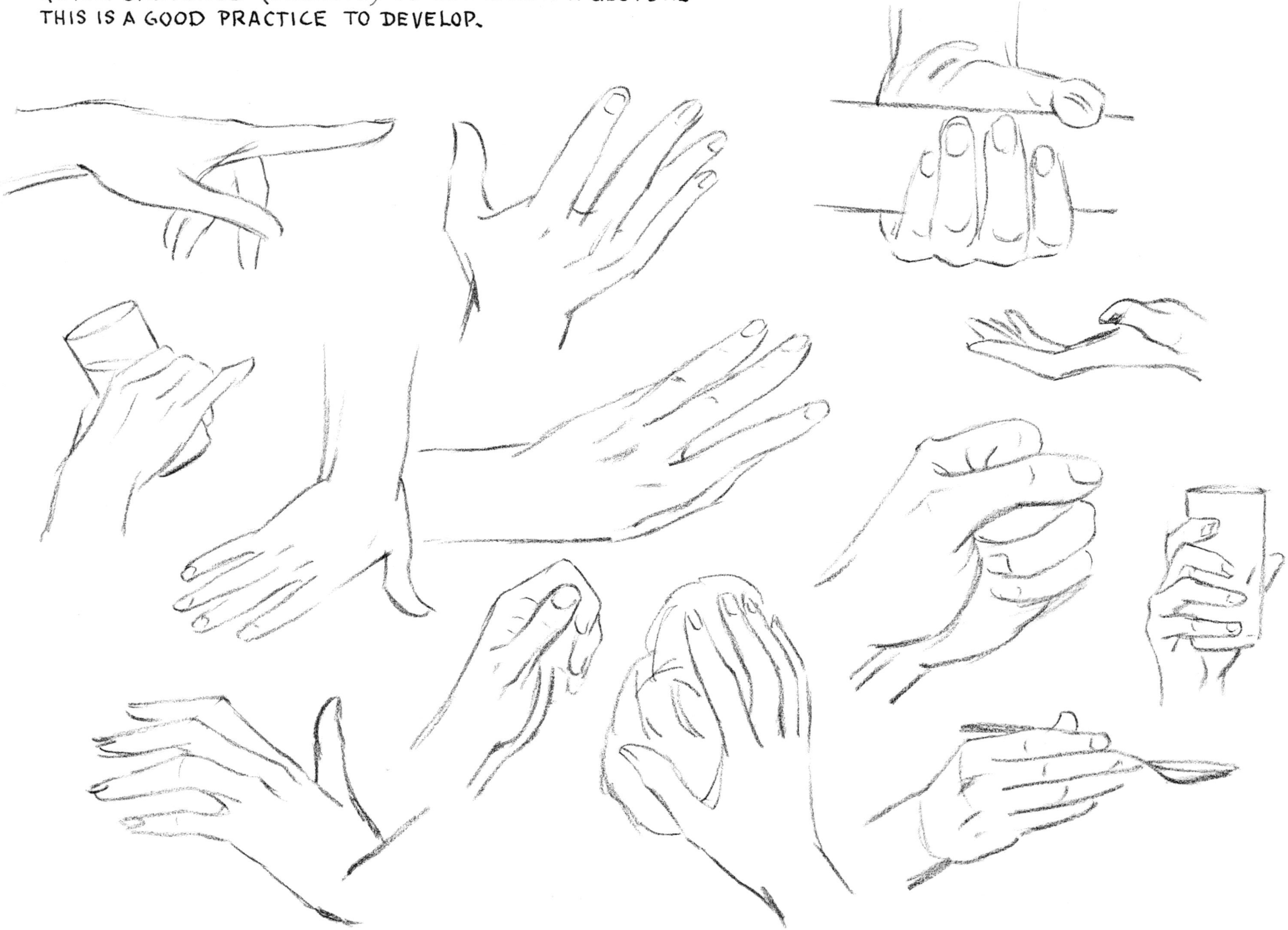

DRAWN ON LIGHT WEIGHT BOND WITH HB WOLFF PENCIL.

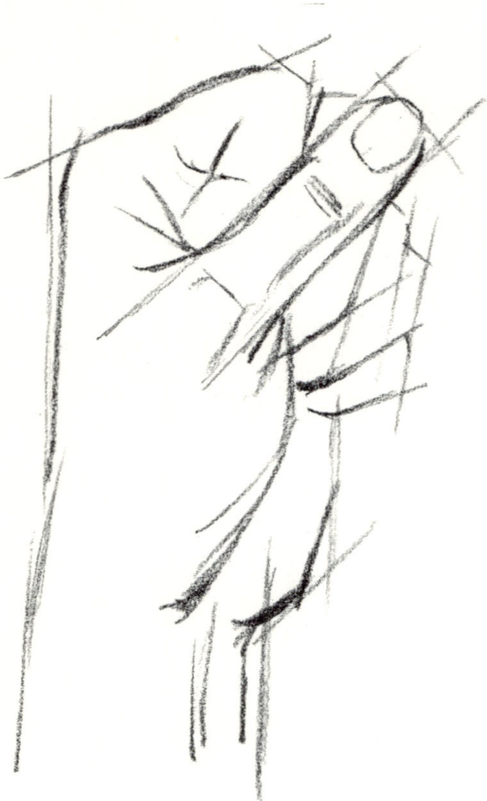

COMPLEX GESTURES CAN BE DRAWN EASIER IF ONE WILL WATCH CAREFULLY THE ANGLES THE VARIOUS PARTS MAKE WITH AN IMAGINARY HORIZONTAL AND VERTICAL REFERENCE LINE. THIS PRINCIPLE IS THE BASIS FOR DRAWING ANY THREE DIMENSIONAL OBJECT ON A TWO DIMENSIONAL SURFACE. IT IS THE PRINCIPLE THAT CREATES THE ILLUSION OF PERSPECTIVE IN DRAWING.

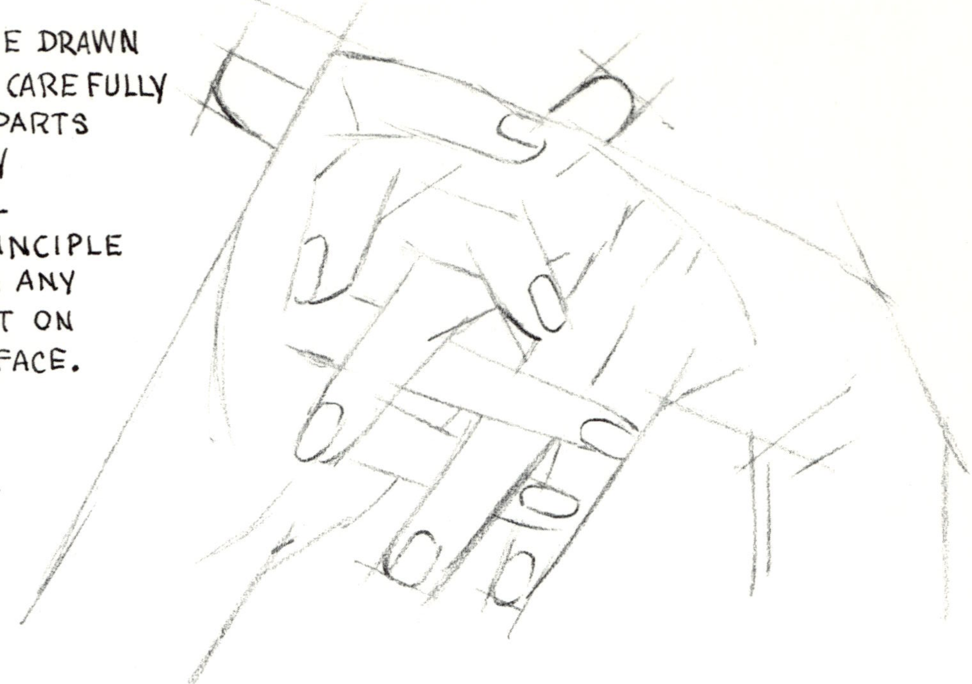

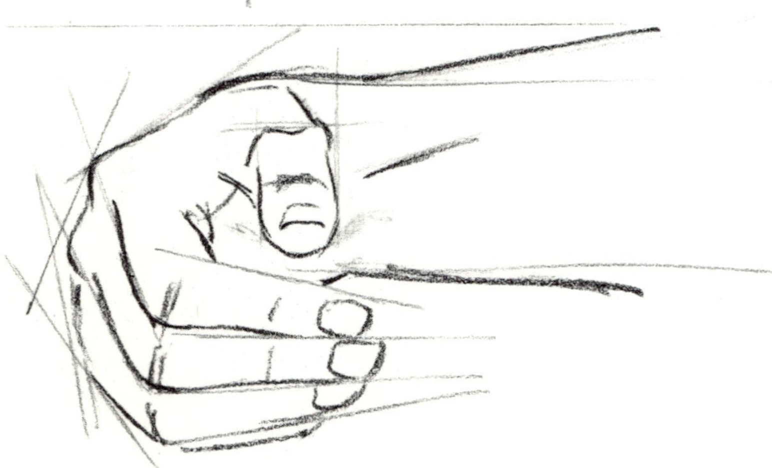

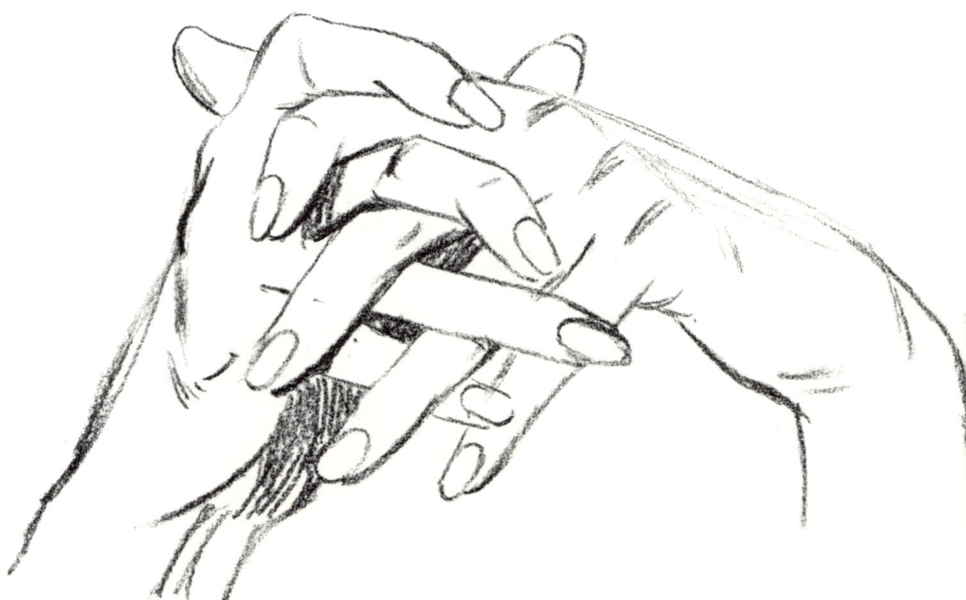

DRAWN ON LIGHT WEIGHT BOND PAPER WITH HB&2B WOLFF PENCILS.

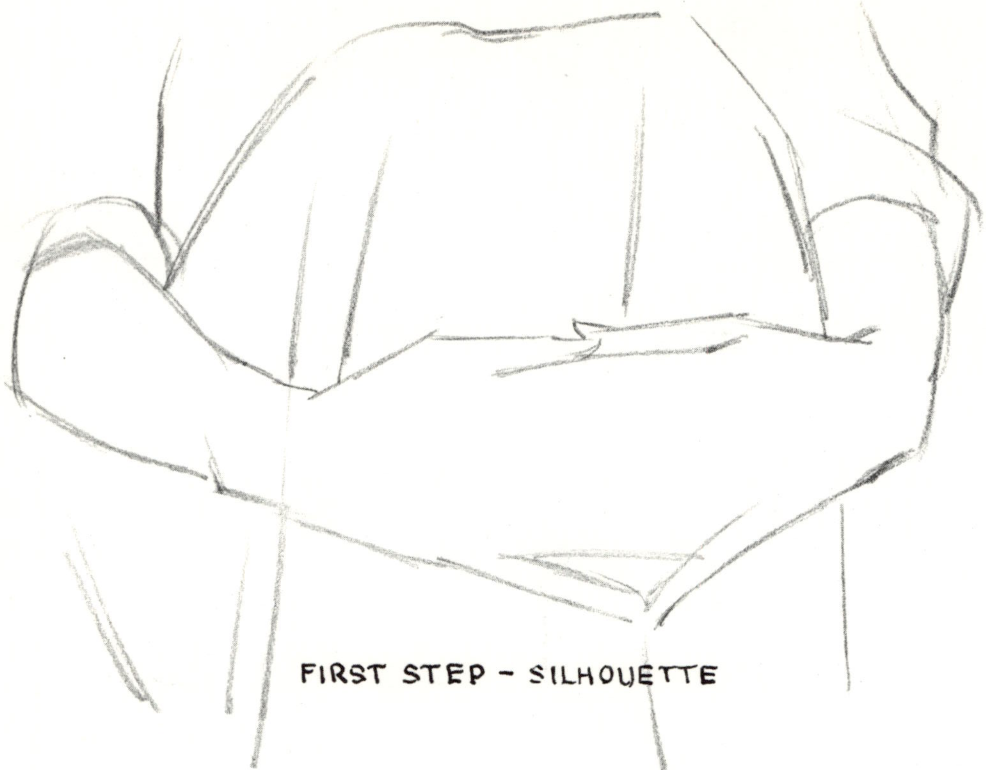

FIRST STEP – SILHOUETTE

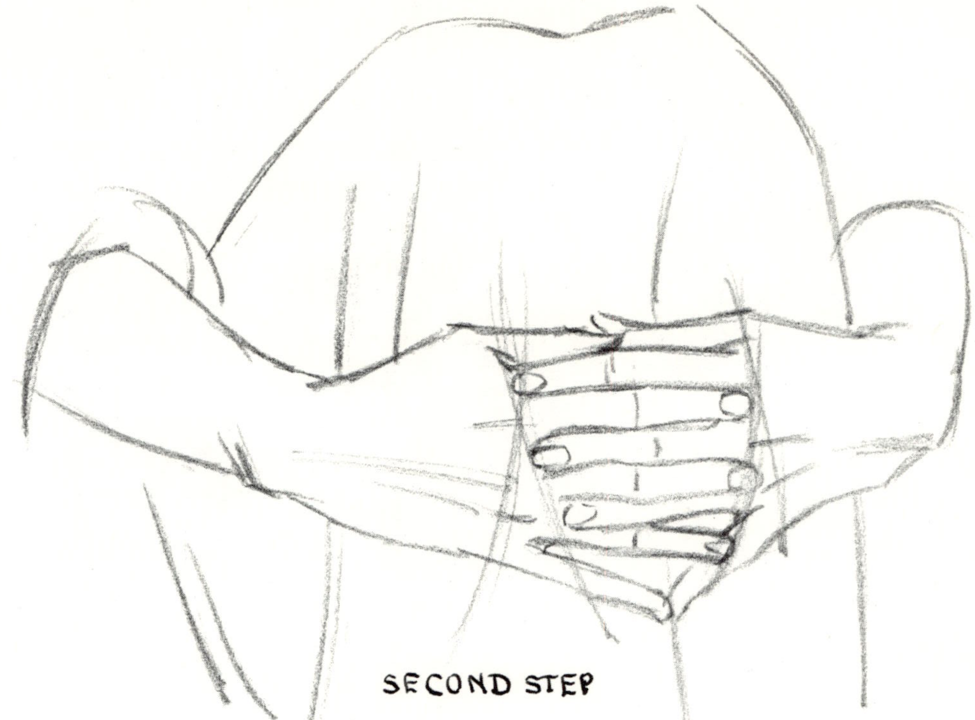

SECOND STEP

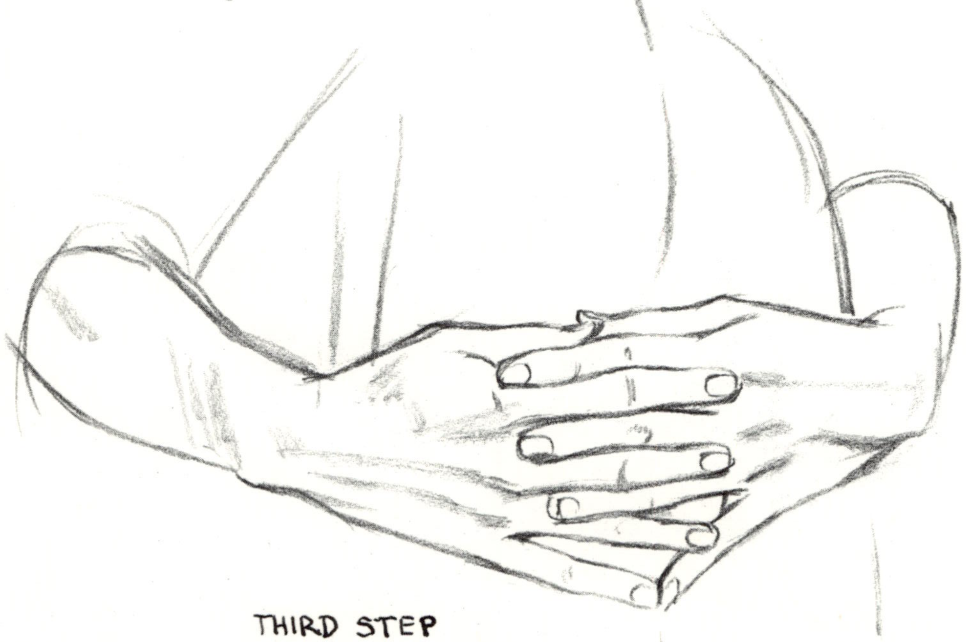

THIRD STEP

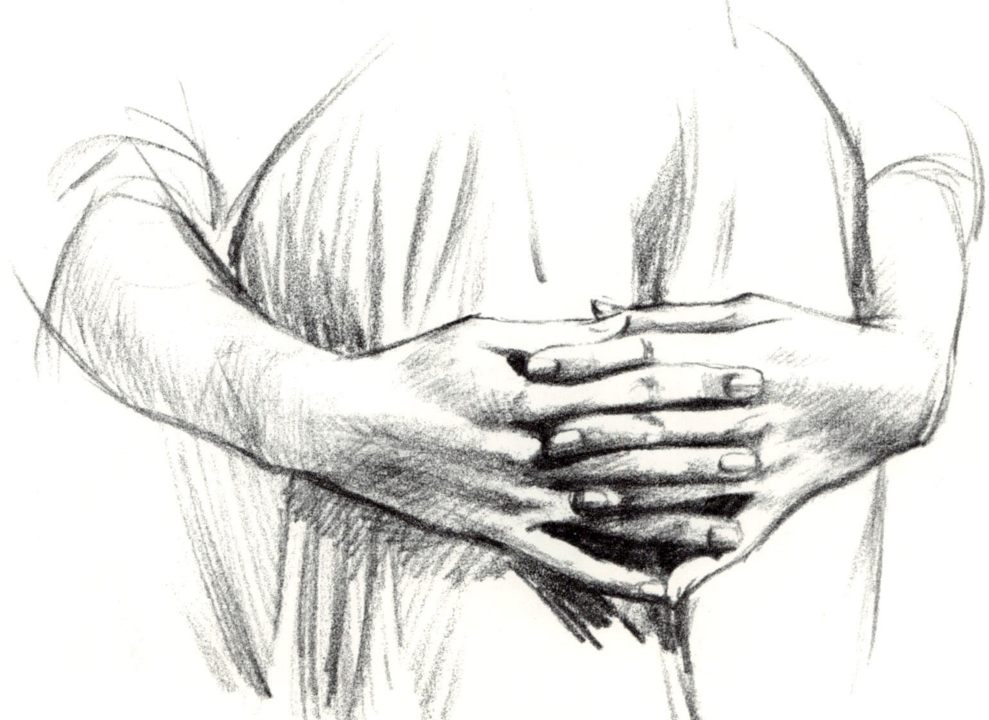

DRAWN ON LIGHT WEIGHT BOND PAPER WITH 2 AND 4B GRAPHITE PENCILS.

DRAWN ON HEAVY ARTIST'S VELLUM.

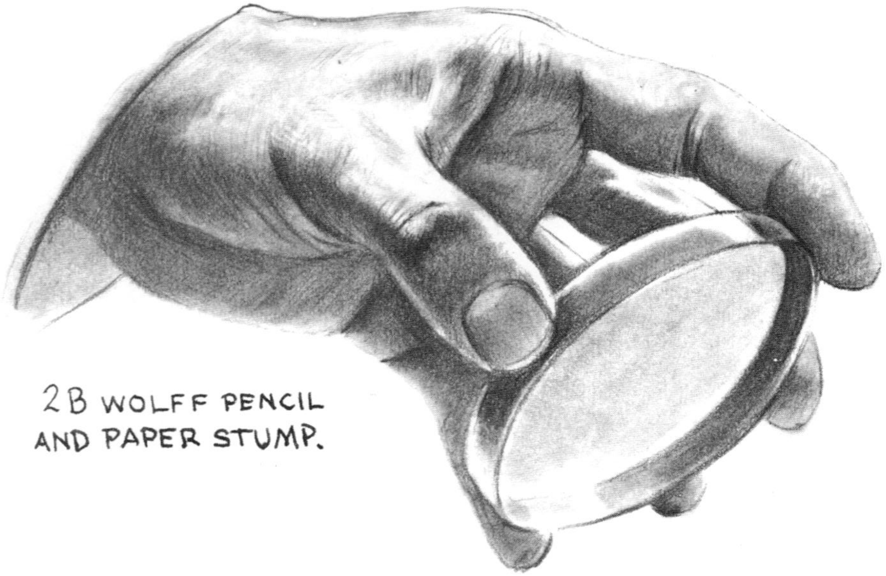

2B WOLFF PENCIL AND PAPER STUMP.

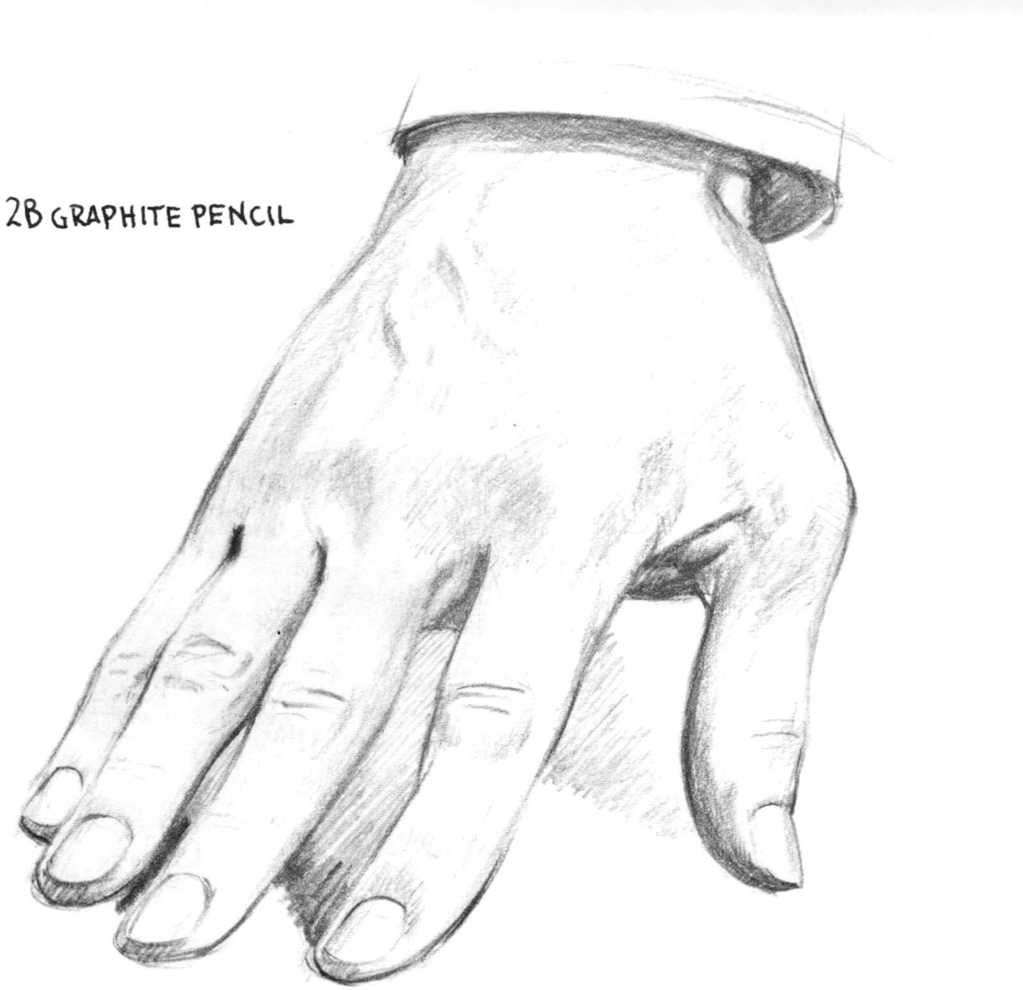

2B GRAPHITE PENCIL

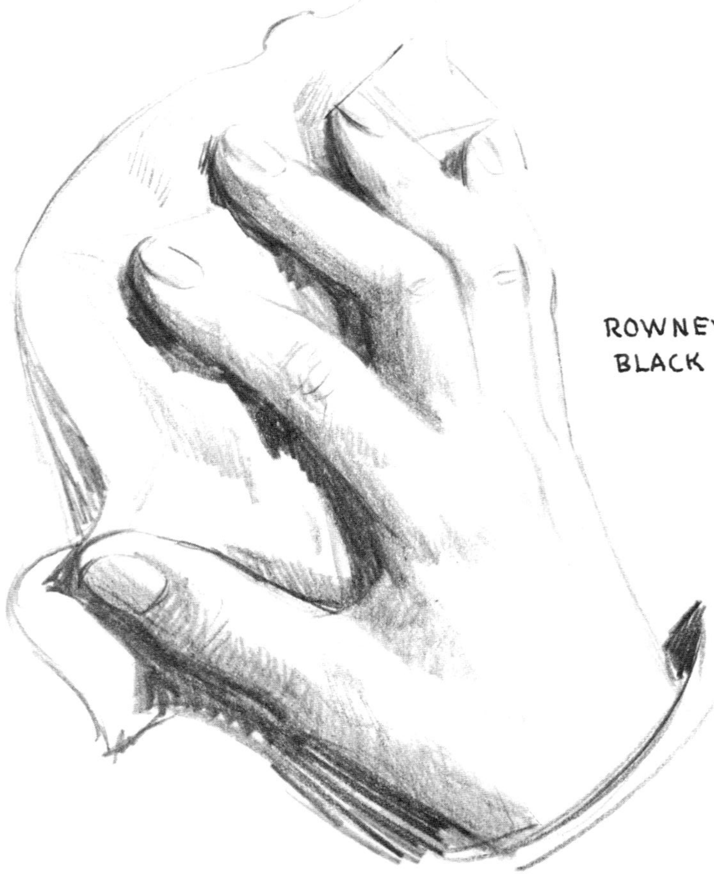

ROWNEY "VICTORIA" BLACK PENCIL.

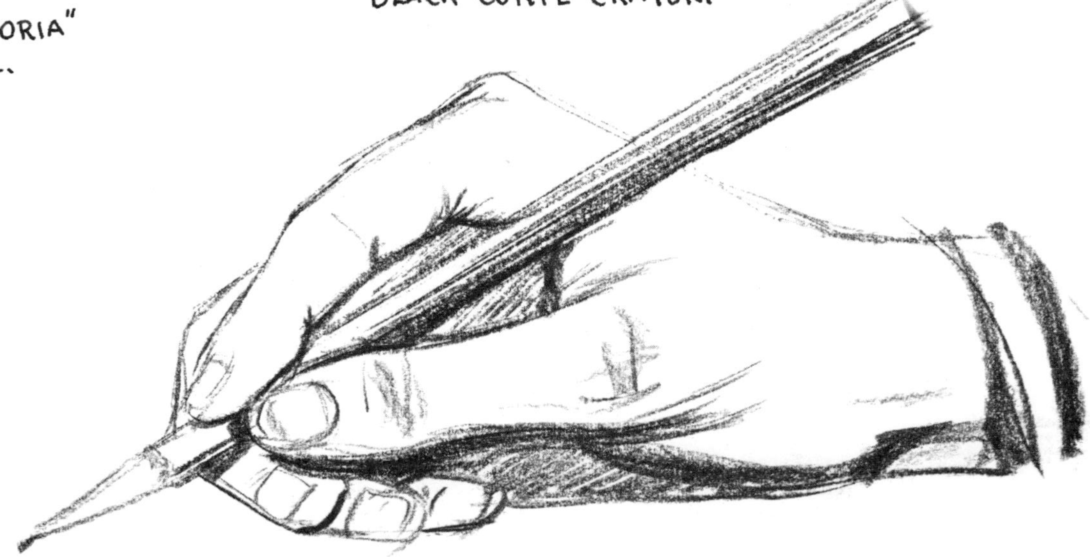

BLACK CONTÉ CRAYON.

DRAWN WITH A STAEDTLER "MARS-700 0.3 PEN, THESE PENS CAN MAKE ONLY A SINGLE WEIGHT LINE. THEY ARE USED MOSTLY FOR DRAWING PLANS AND SINGLE STROKE LETTERING

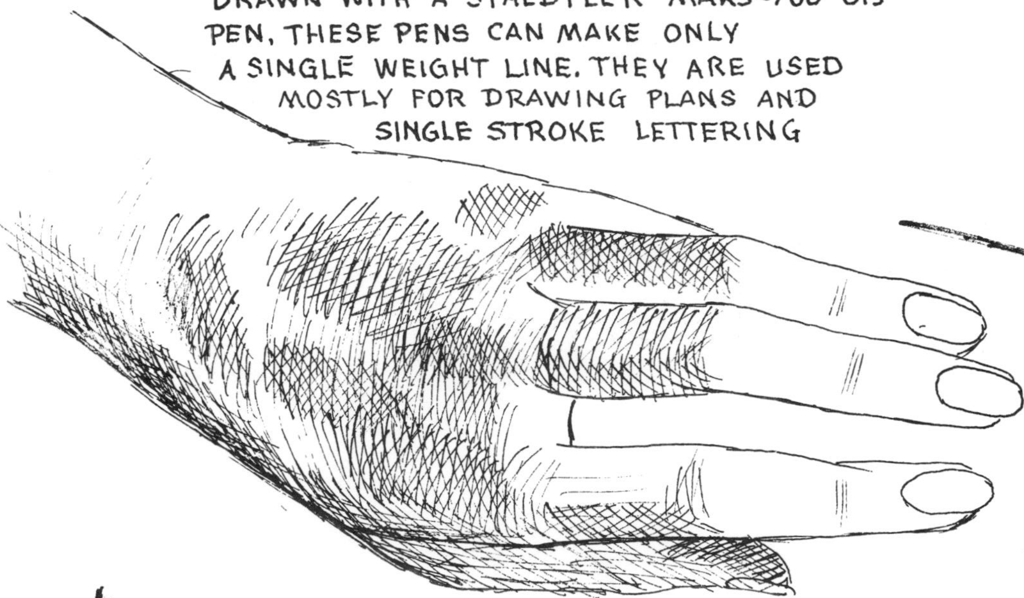

PENTEL PEN

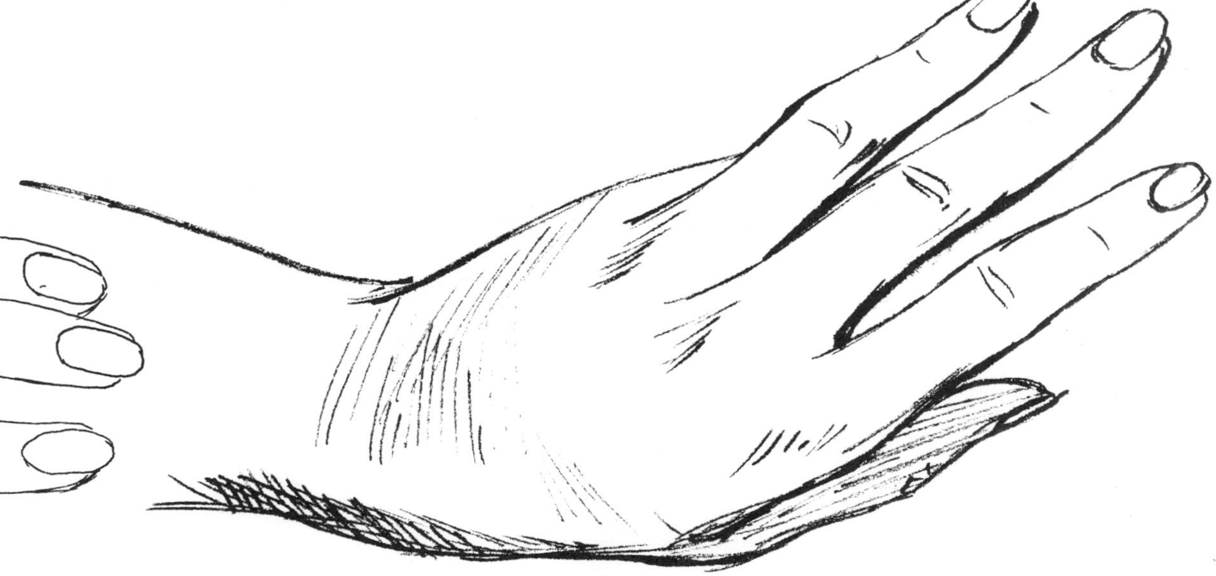

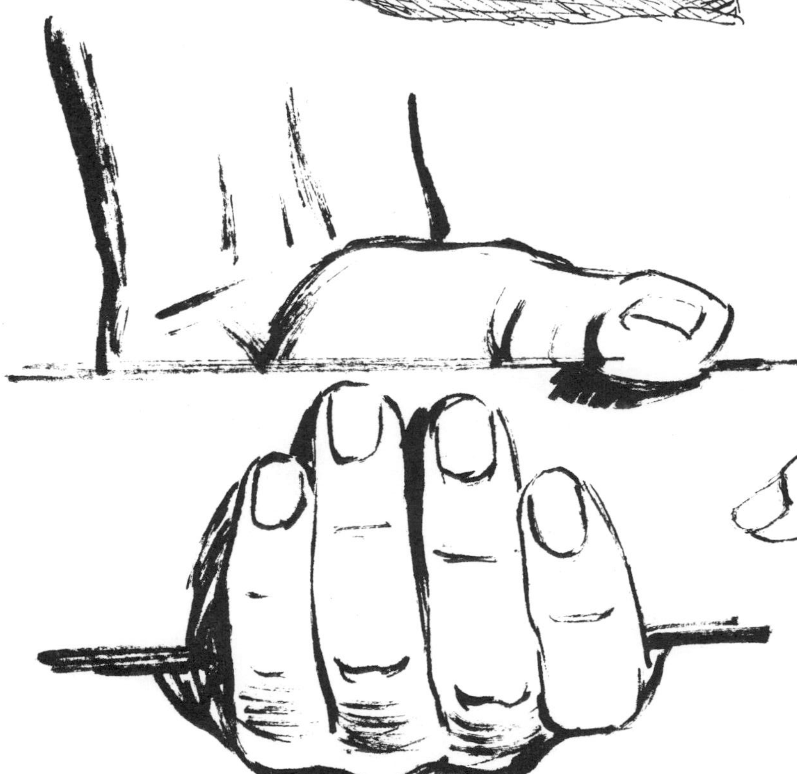

FLO-MASTER FELT TIP PEN.

DRAWN ON MEDIUM WEIGHT BOND PAPER.

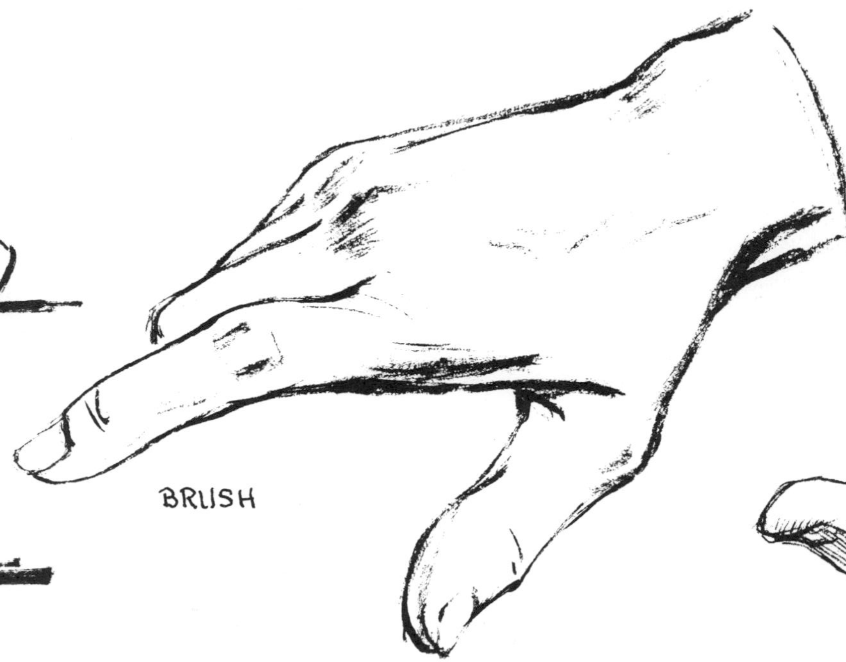

BRUSH

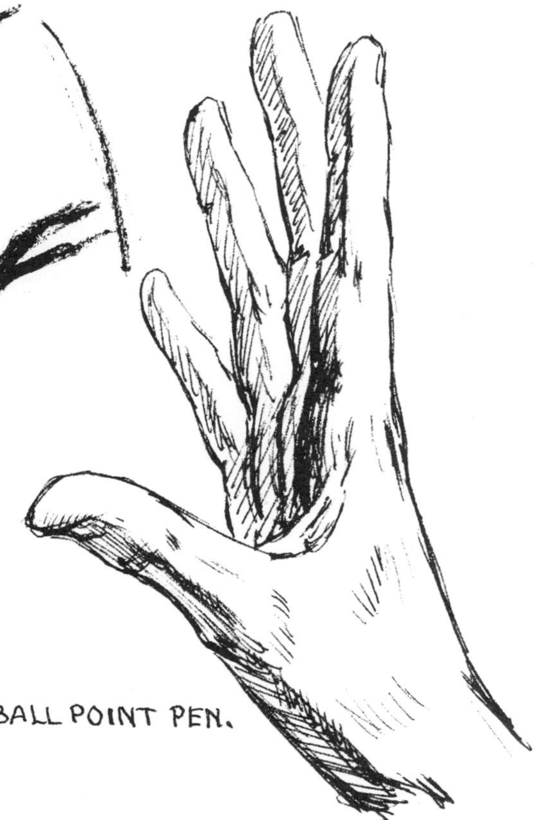

COMMON BALL POINT PEN.

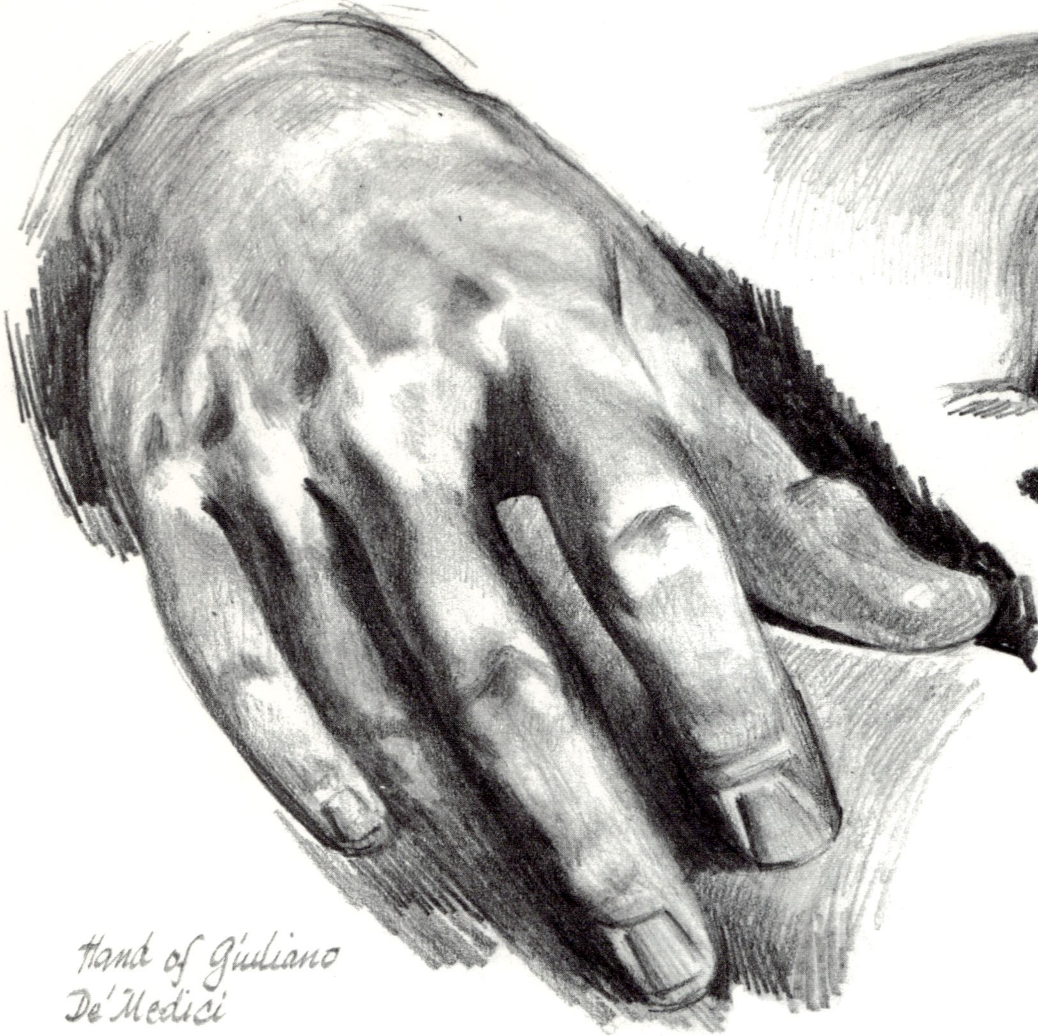

Hand of Giuliano De' Medici

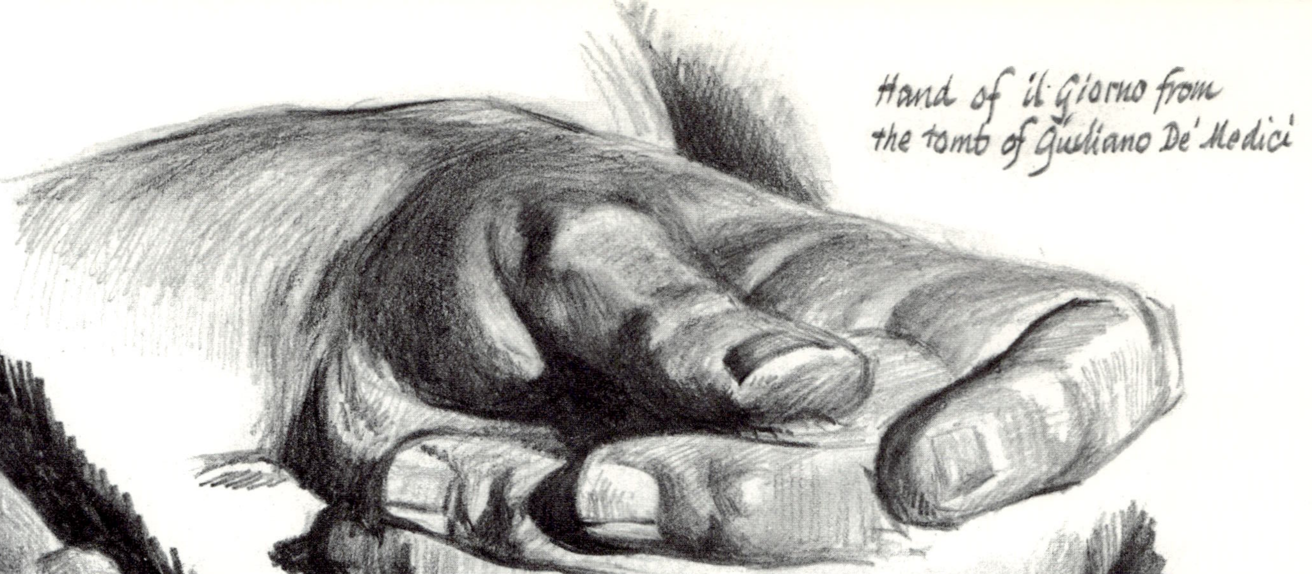

Hand of il Giorno from the tomb of Giuliano De' Medici

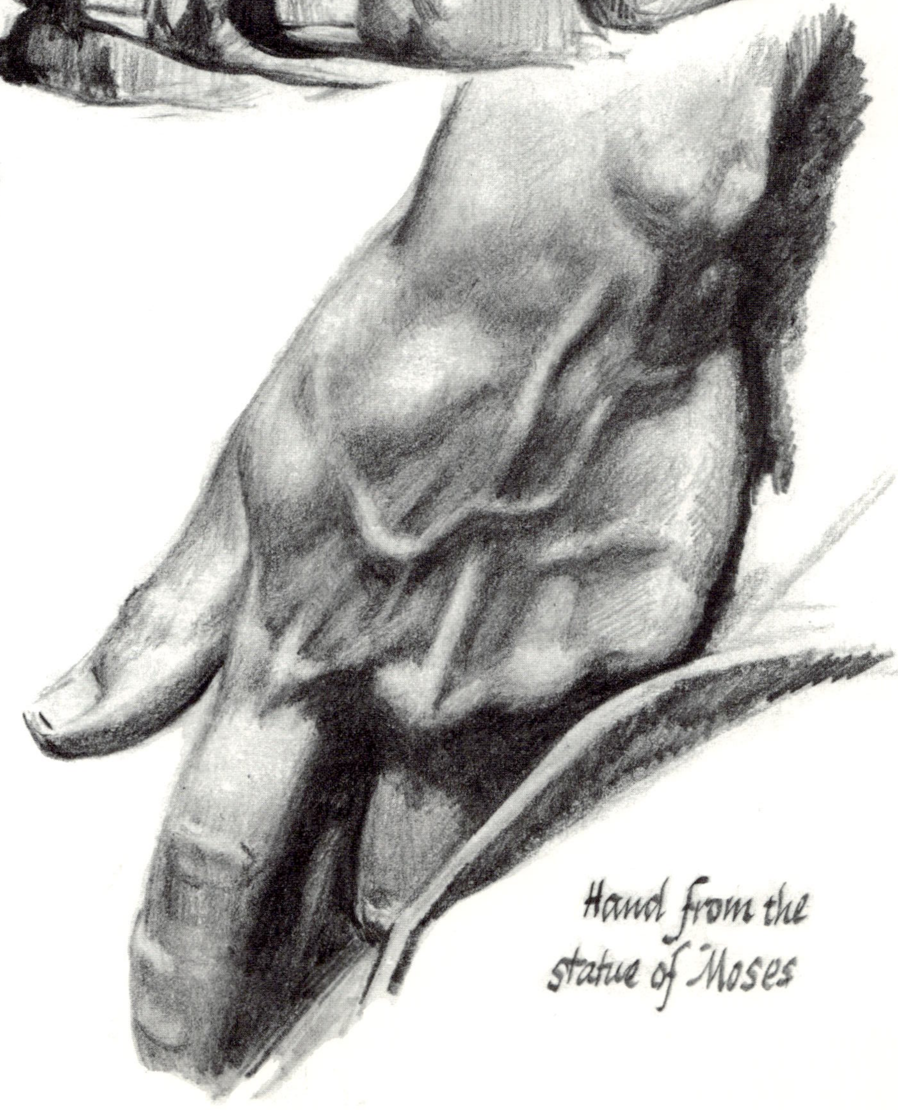

Hand from the statue of Moses

These drawings were made from photographs of Michelangelo's sculpture. Michelangelo was one of the great masters of form. His works in sculpture are worthy of close study by all who are interested in portraying human form.

ACTION STUDIES OF THE HAND

THE FOLLOWING portfolio of drawings has been done with various dry and wet mediums. The postures chosen have been the most commonplace of gestures. Some have been drawn from photographs taken by me, while others have been drawn directly from the model. Several have been drawn from the mirrored reflection of my own hand. One need not protest that there are no models to draw from since our news magazines and newspapers are full of pictures of hands in action. There are also members of the family as well as friends who will gladly pose for the serious student.

Drawing is an empirical art. One may read constantly from books dealing with drawing, yet never learn to draw unless one actively engages in the act. Competence can be gained solely from actual drawing — manual dexterity is achieved only by the actual doing. Through constant observation and physically drawing, one gains the ability to draw with conviction. Study of the drawings, or at least the reproductions, of the masters of drawing, old and contemporary, can be most rewarding.

Drawing is the foundation of all graphic efforts. Even writing by hand is drawing — it is, or should be, the clear drawing of an abstract symbol, denoting a sound or meaning. Unreadable writing is akin to mumbled speech. Unintelligible drawing is in the same category. That which is well drawn is clear to all. Drawing is a universal language, understood even by the aborigines. Drawing is the graphic statement of things observed and understood. This does not mean that a drawing must contain all the details of an object, stated down to the most miniscule observation. Individual character develops in drawing, in the same manner that we develop an accent in speech, unconsciously. A conscious effort to develop a technique in drawing always results in a thin gimmicky group of tricks that become tiresome to the observer.

Drawing and painting are synonymous — only the mediums are different. We have a tendency to separate drawing and painting. Charles W. Hawthorne, a great painter and a great teacher, said this about painting, "Painting is really quite easy — all it is, is to get the right color and the right value and put it in the right place." That, of course, is true, but an oversimplified statement. We could paraphrase his statement by saying, "Drawing is quite easy, all one has to do is put the line and the tonal value in the right place." The latter statement can be achieved only through constant practice.

I suggest, no, I urge you, to initiate a program of drawing and to maintain it on a daily basis. A program of this nature, I can assure you, shall have you achieving goals beyond your fondest hopes.

DRAWN WITH HB WOLFF PENCIL

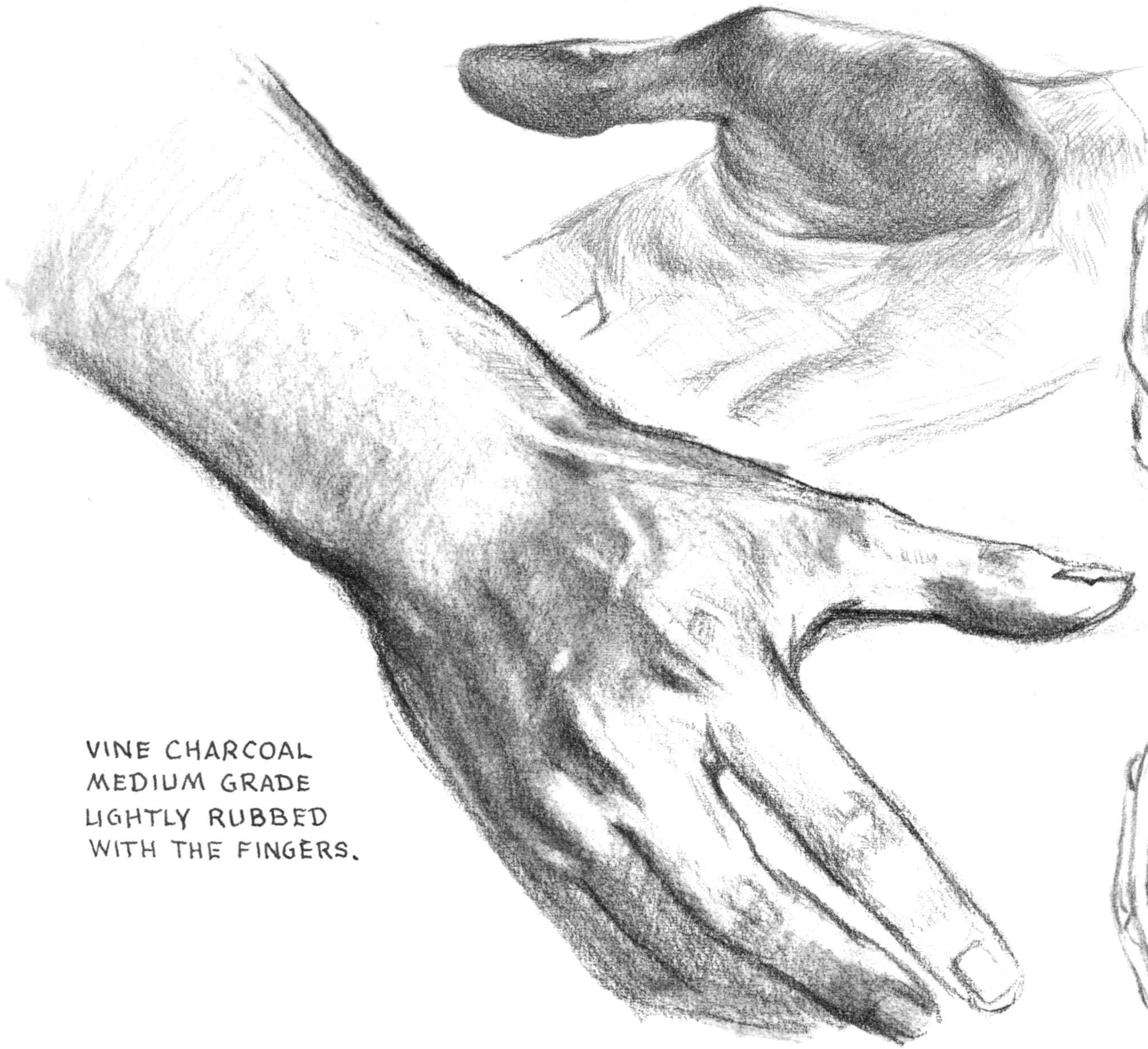

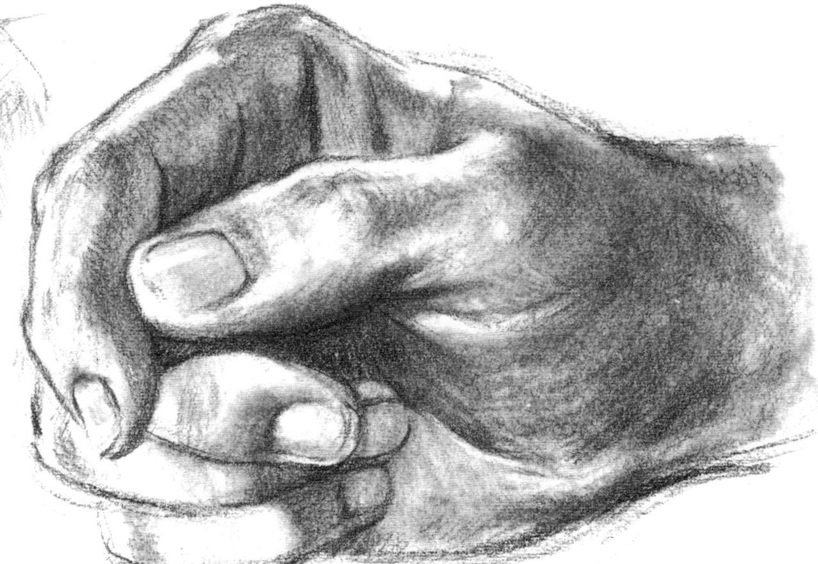

EAGLE "CHARCO" 2B AND PAPER STUMP.

VINE CHARCOAL
MEDIUM GRADE
LIGHTLY RUBBED
WITH THE FINGERS.

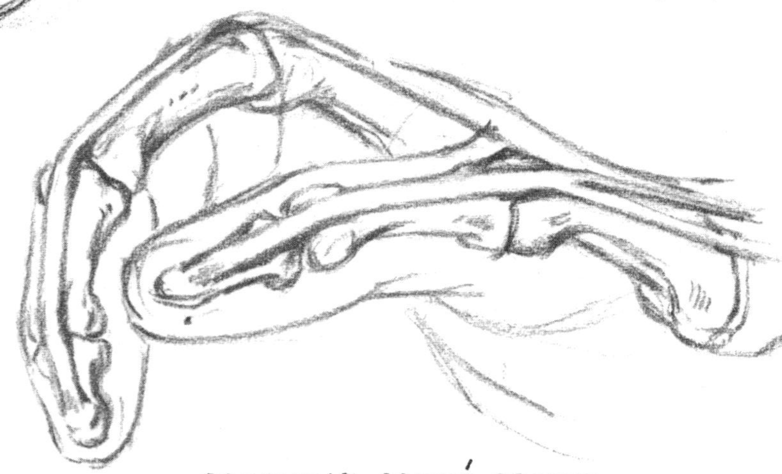

SANGUINE CONTÉ CRAYON.

DRAWN ON HEAVY WEIGHT BOND PAPER.

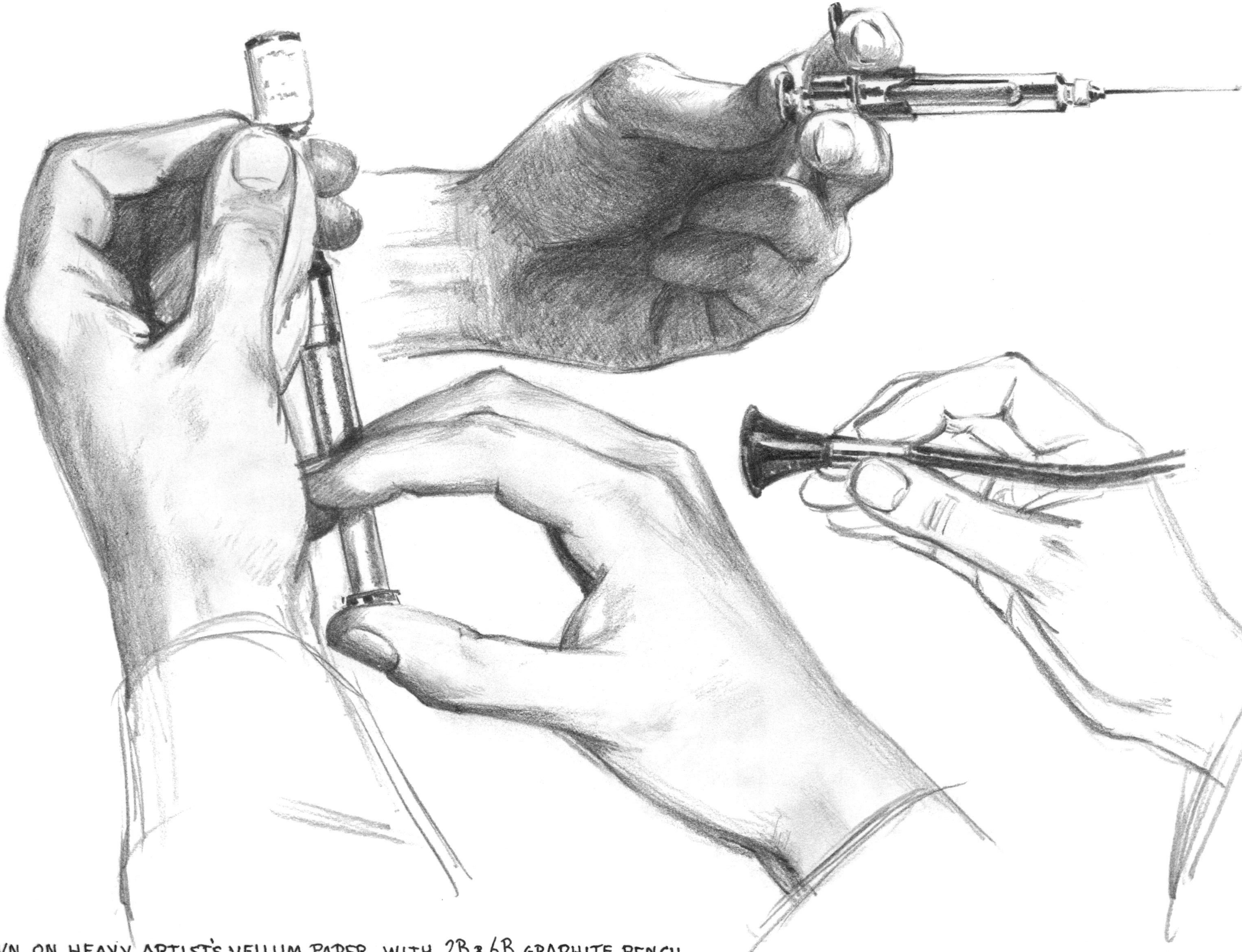

DRAWN ON HEAVY ARTIST'S VELLUM PAPER WITH 2B & 6B GRAPHITE PENCIL.

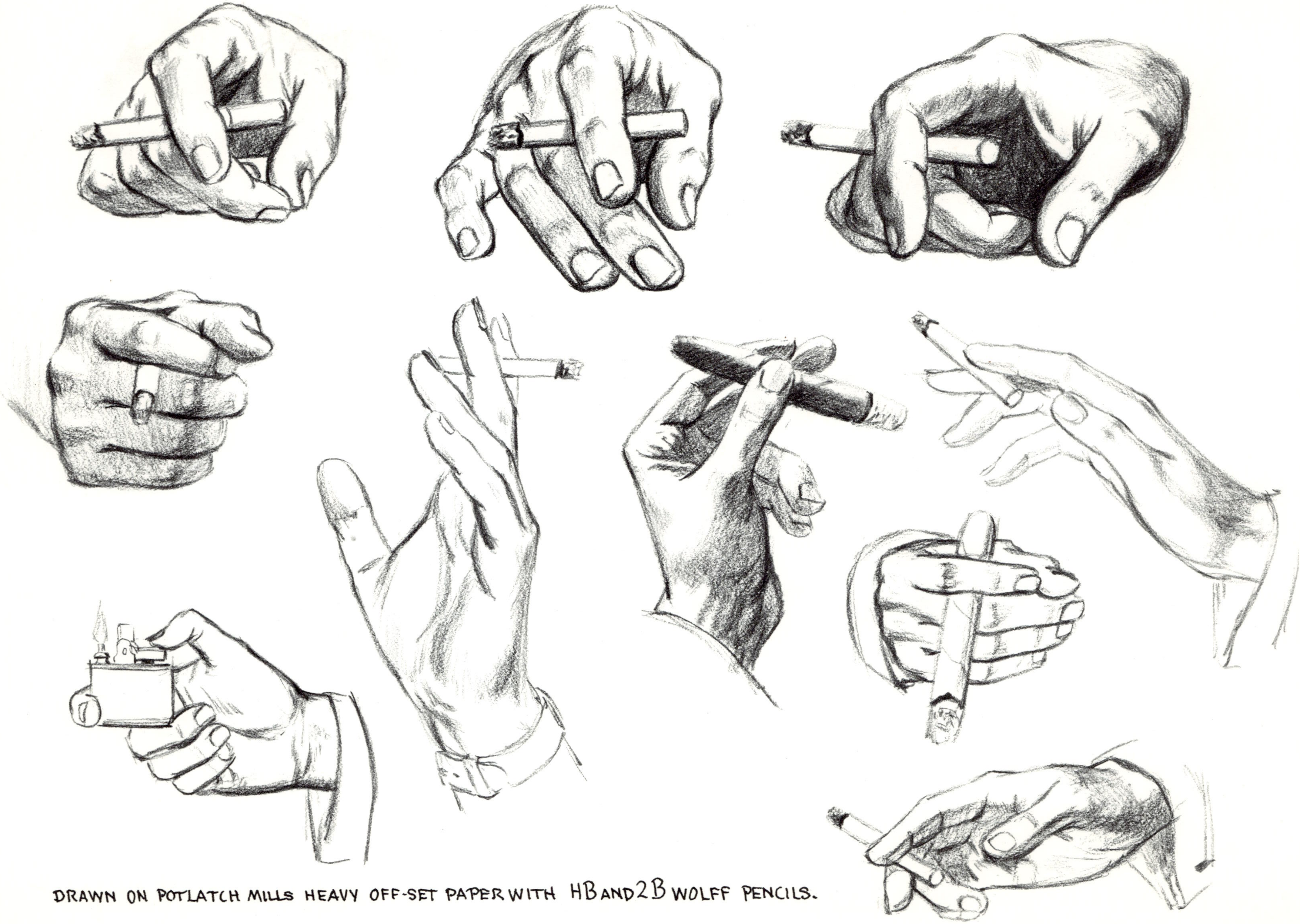

DRAWN ON POTLATCH MILLS HEAVY OFF-SET PAPER WITH HB AND 2B WOLFF PENCILS.

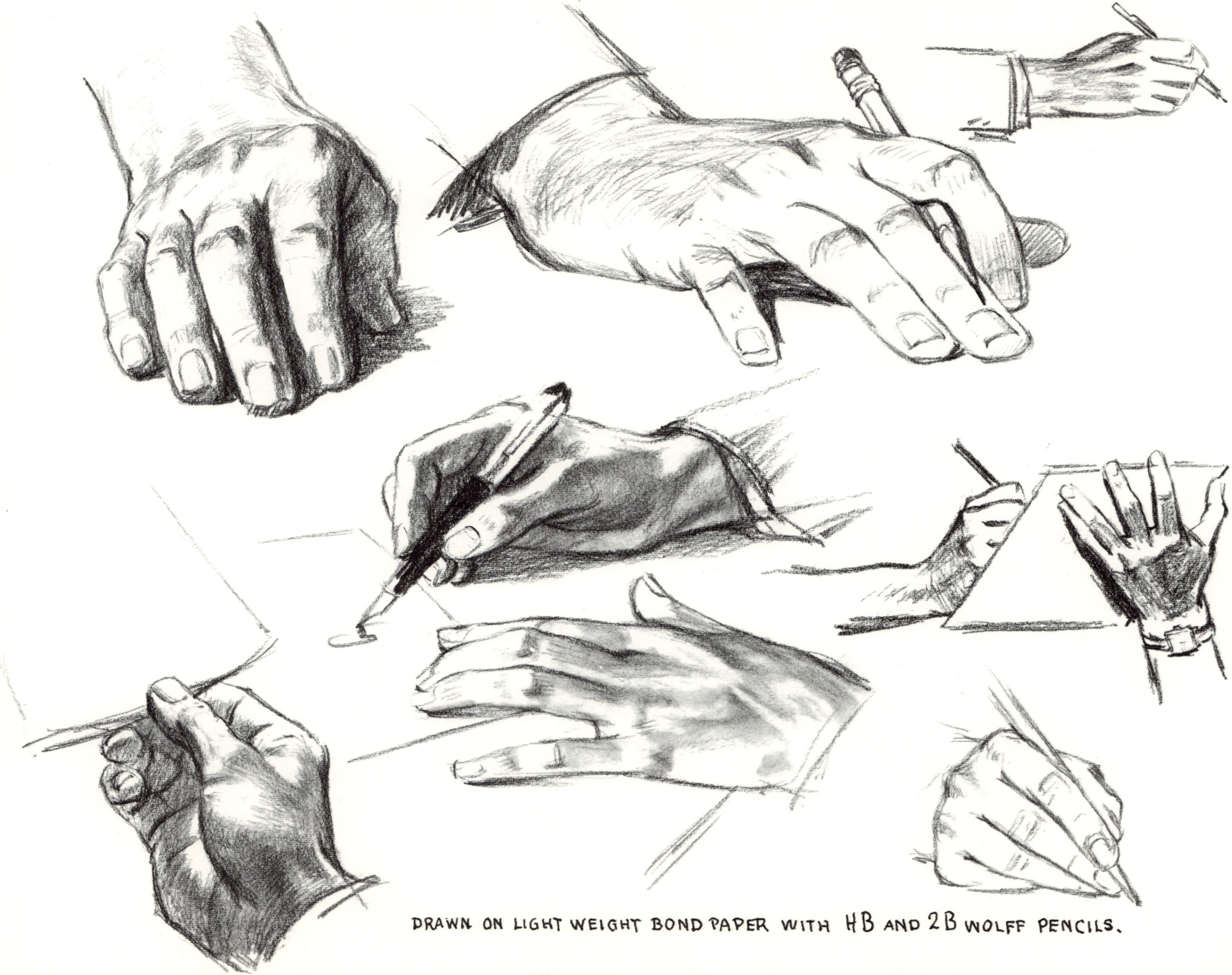

DRAWN ON LIGHT WEIGHT BOND PAPER WITH HB AND 2B WOLFF PENCILS.

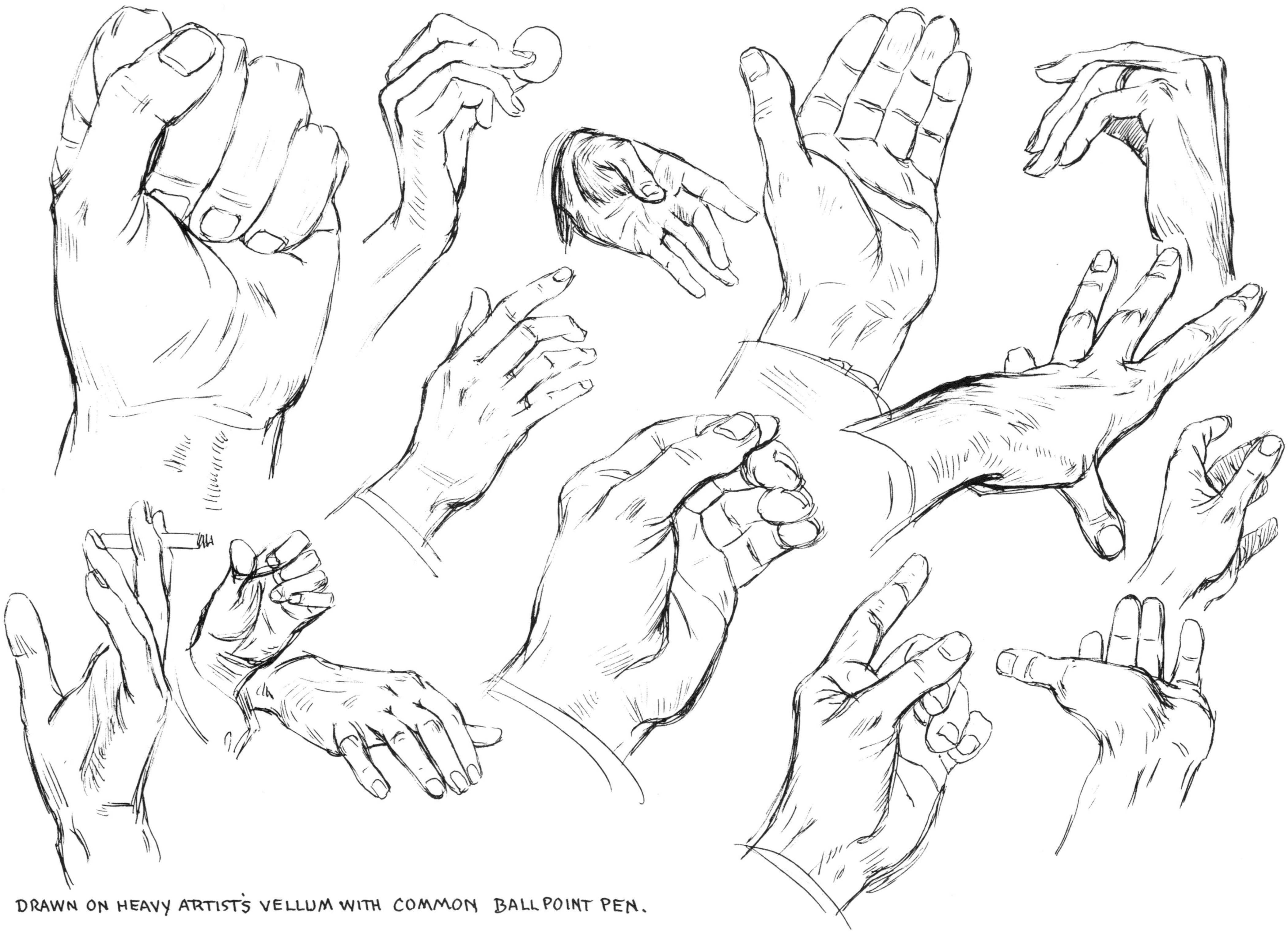

DRAWN ON HEAVY ARTIST'S VELLUM WITH COMMON BALLPOINT PEN.

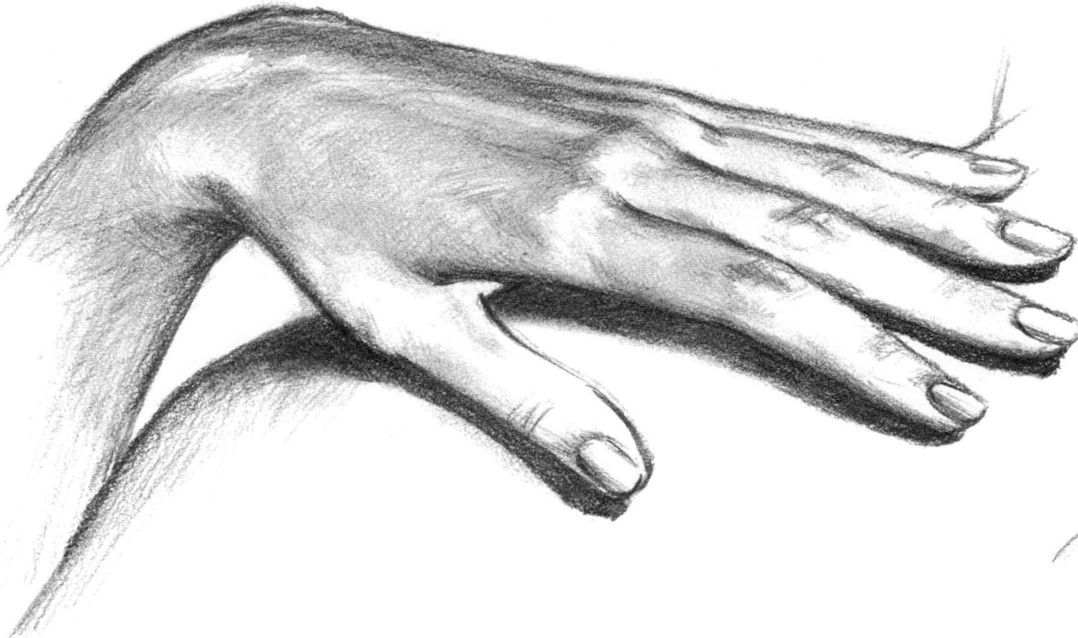

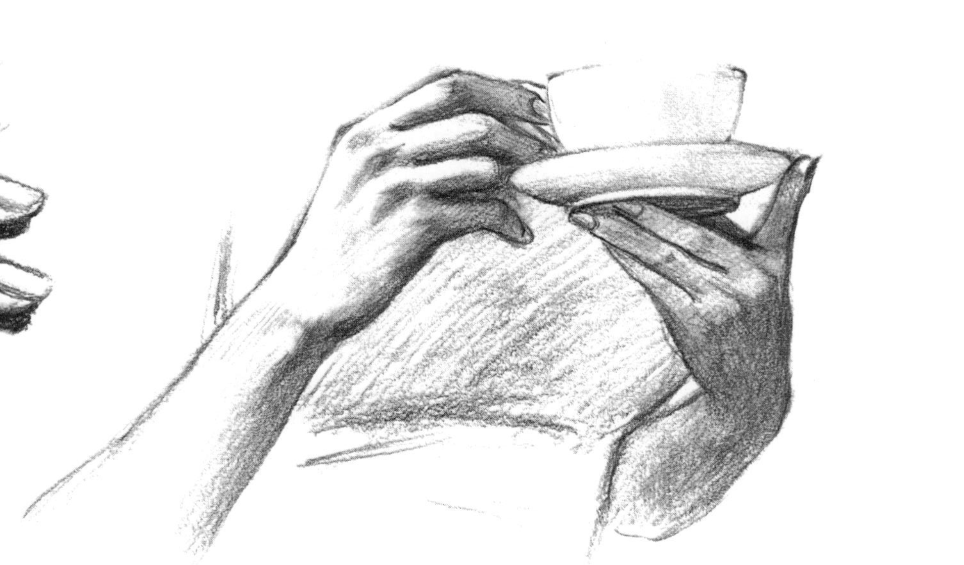

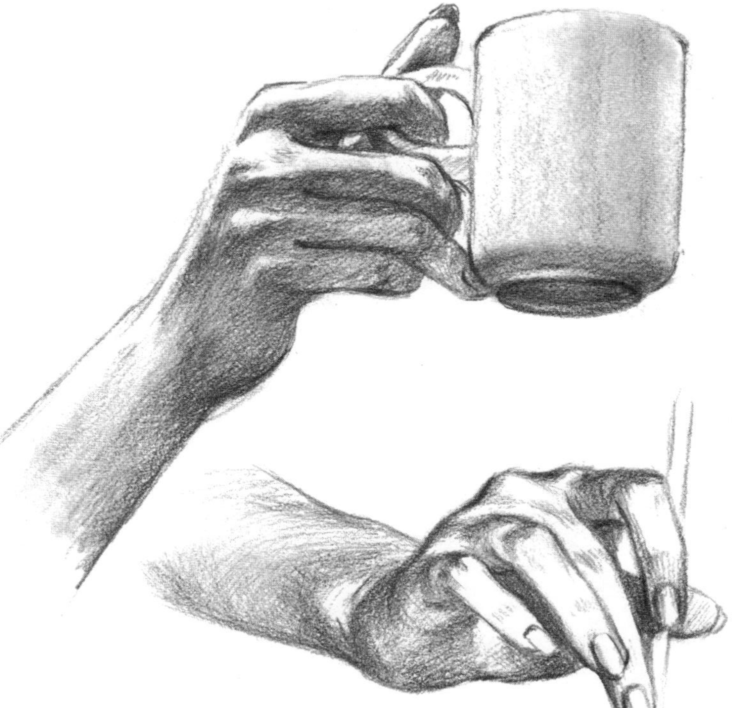

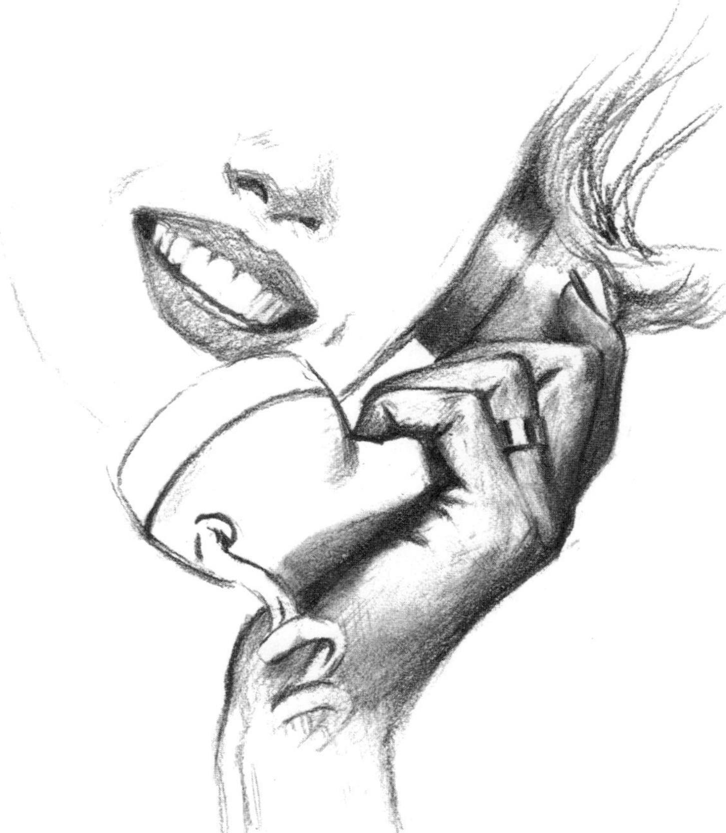

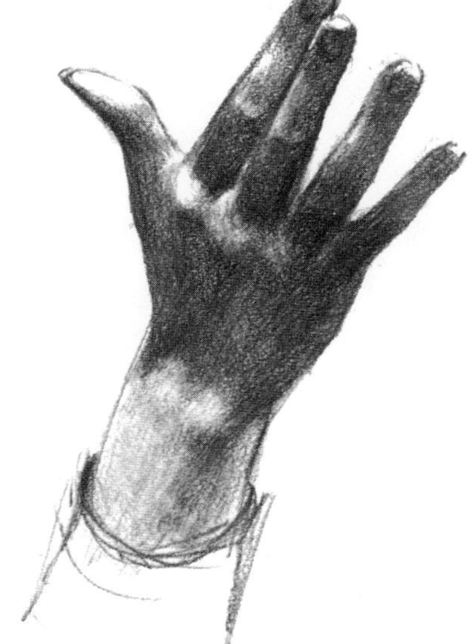

DRAWN ON POTLATCH MILLS HEAVY OFF-SET
PAPER WITH HB, 2B AND 4B WOLFF PENCILS
AND PAPER STUMP.

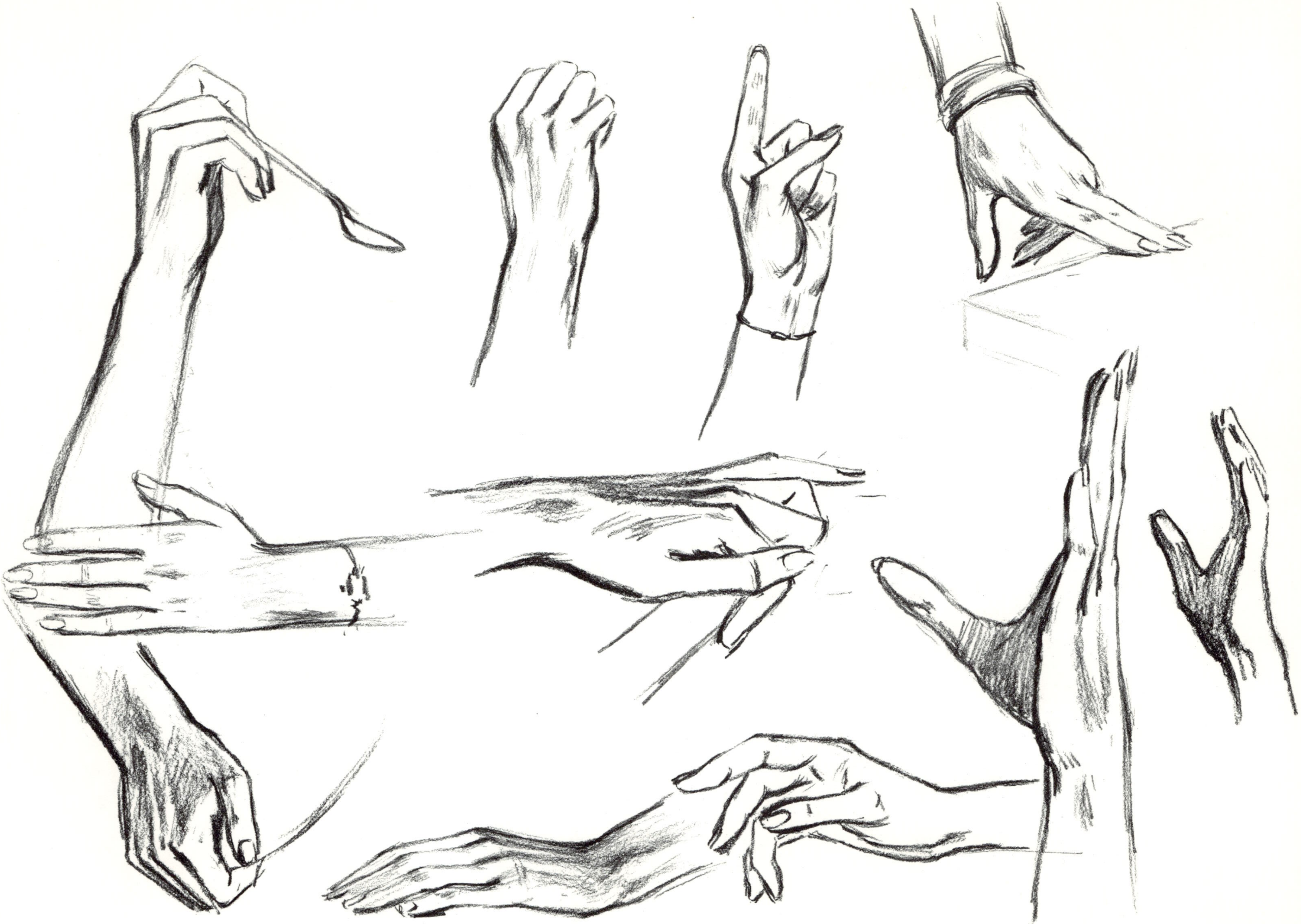

DRAWN ON POTLATCH MILLS HEAVY OFF-SET PAPER WITH B WOLFF PENCIL.

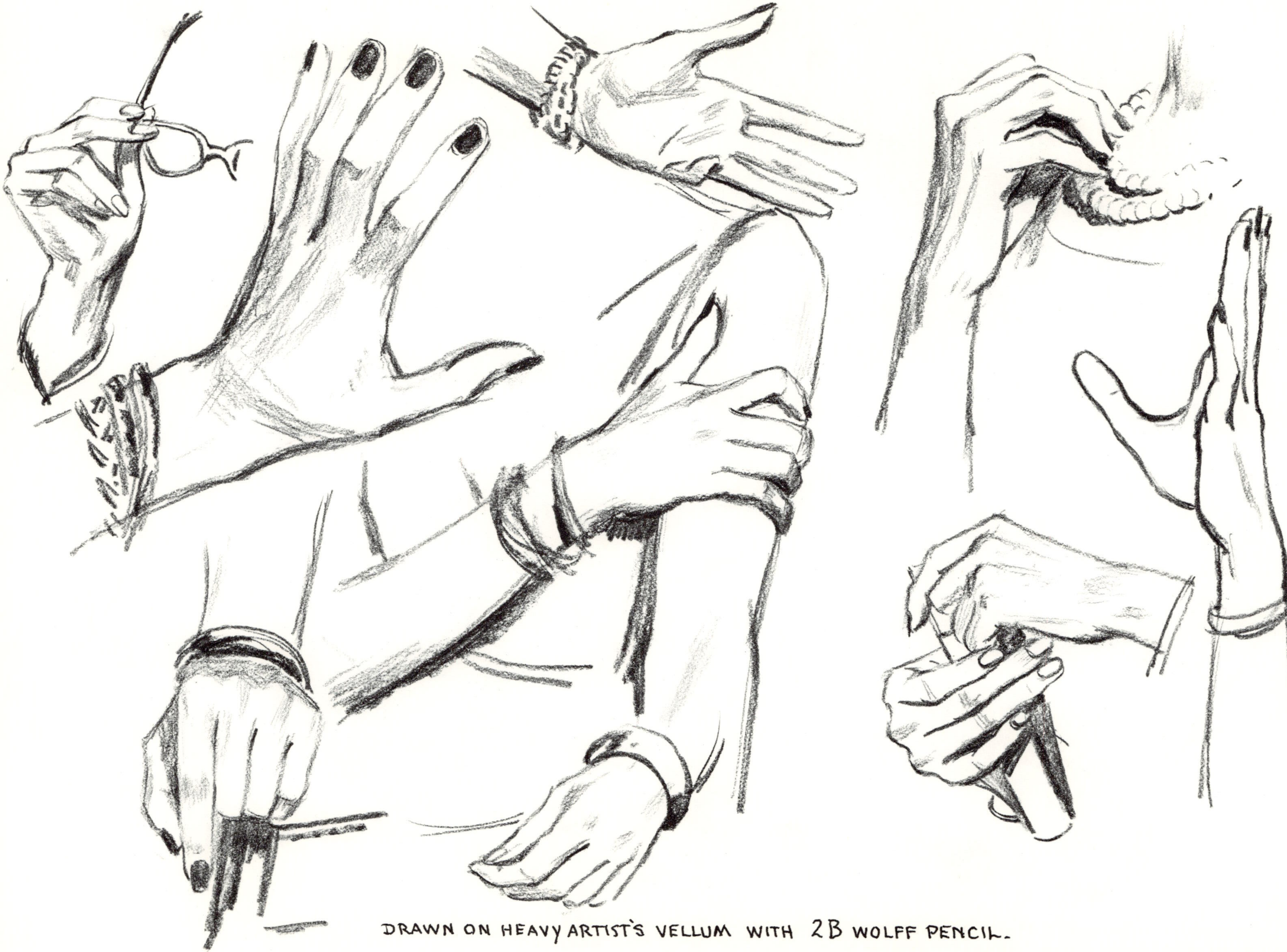

DRAWN ON HEAVY ARTIST'S VELLUM WITH 2B WOLFF PENCIL.

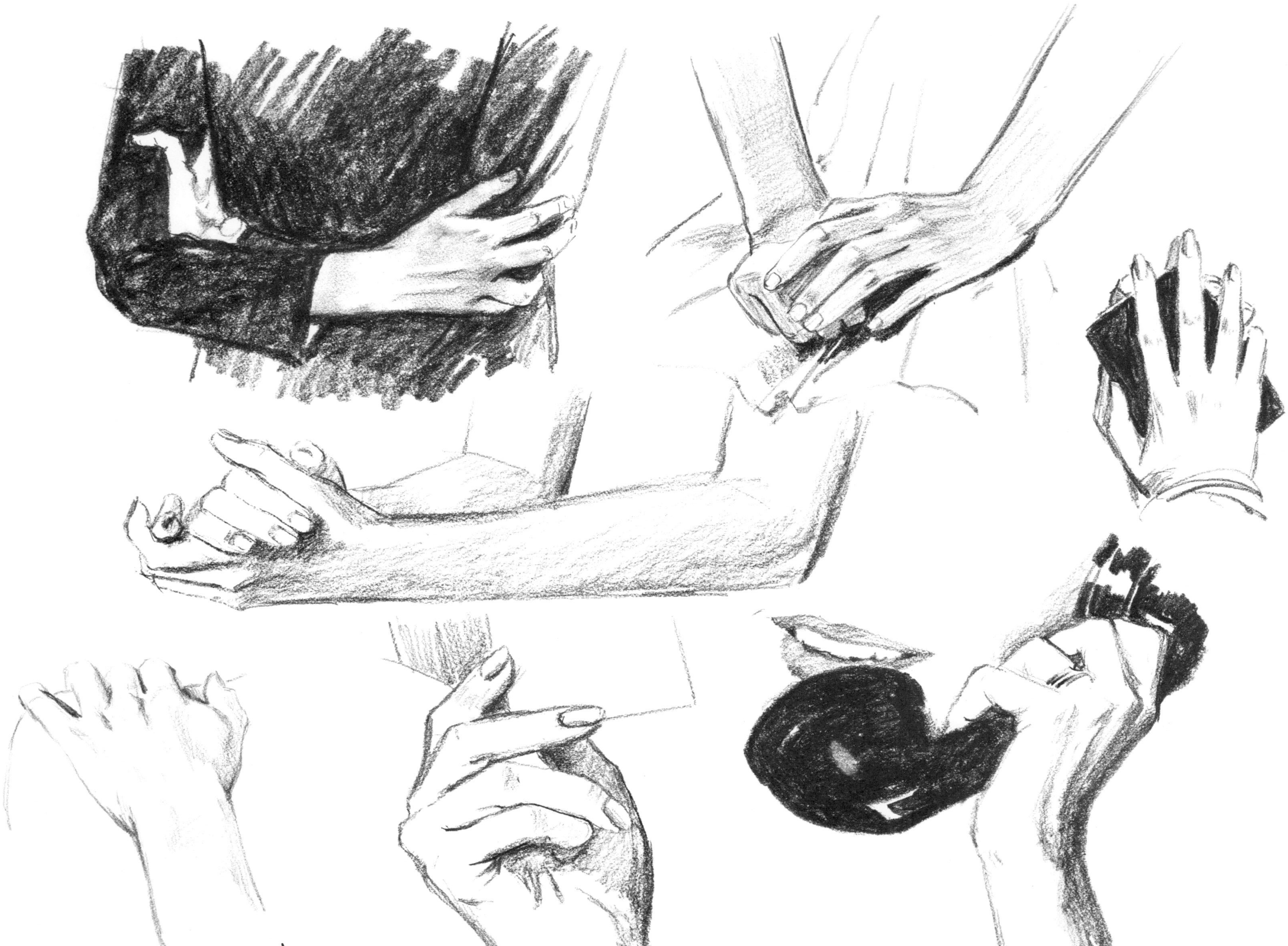

DRAWN ON HEAVY ARTIST'S VELLUM WITH EAGLE HB, 2B AND 6B "CHARCO" PENCILS.

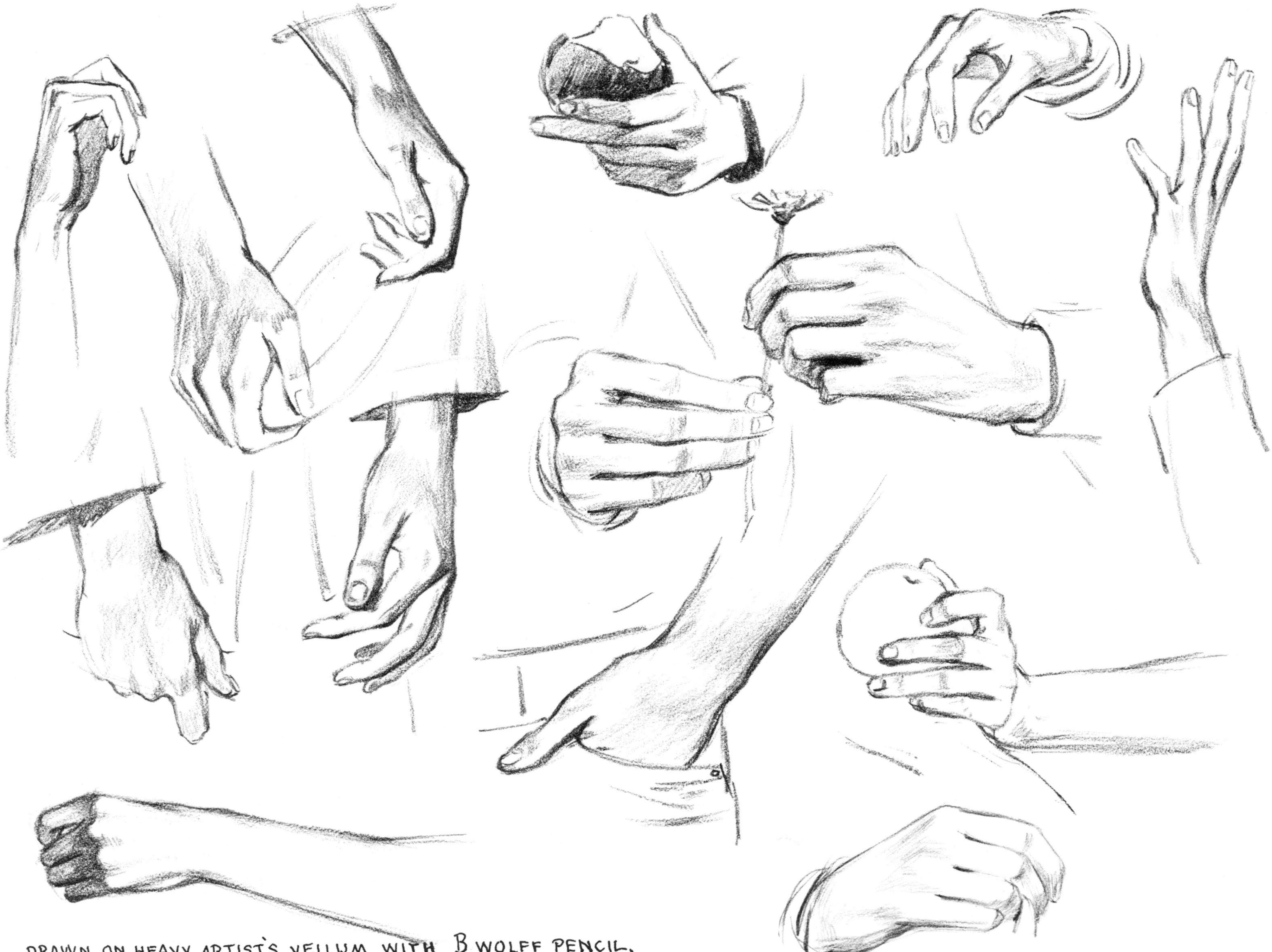

DRAWN ON HEAVY ARTIST'S VELLUM WITH B WOLFF PENCIL.

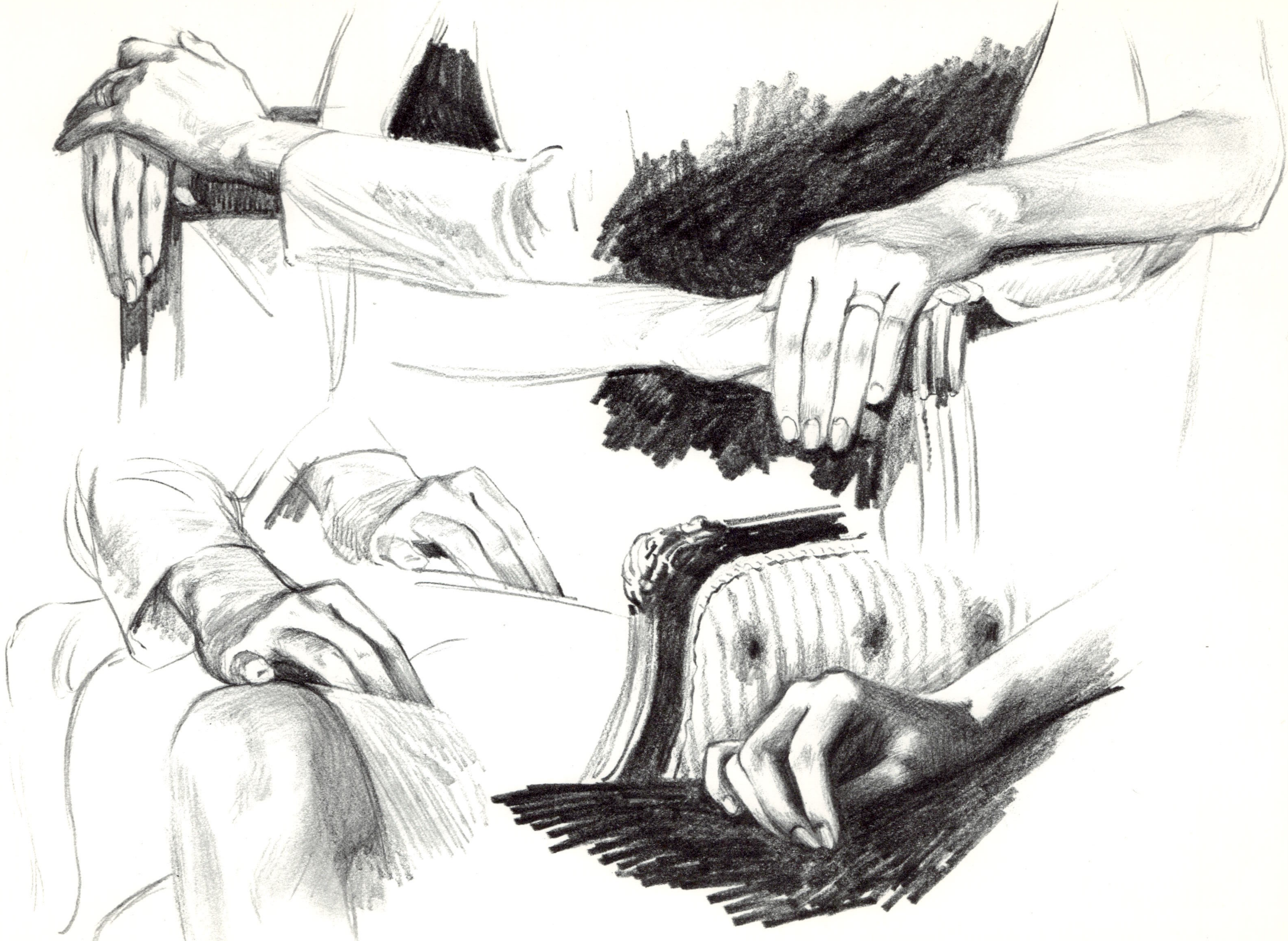

DRAWN ON HEAVY ARTIST'S VELLUM WITH 2B AND 6B GRAPHITE PENCILS.

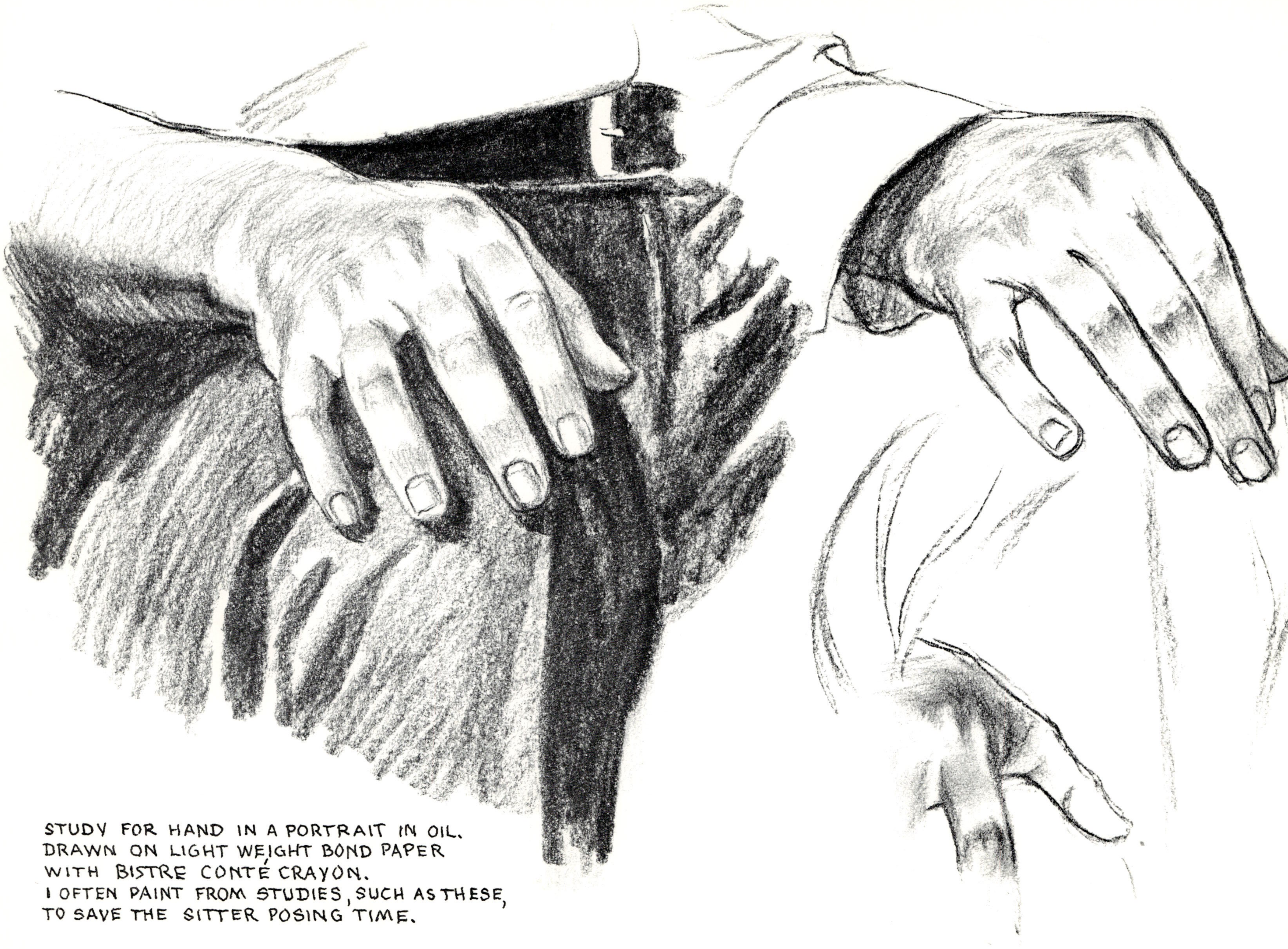

STUDY FOR HAND IN A PORTRAIT IN OIL.
DRAWN ON LIGHT WEIGHT BOND PAPER
WITH BISTRE CONTÉ CRAYON.
I OFTEN PAINT FROM STUDIES, SUCH AS THESE,
TO SAVE THE SITTER POSING TIME.

STUDIES OF HANDS FOR PORTRAITS - PAINTED ON CANVAS WITH ACRYLICS.

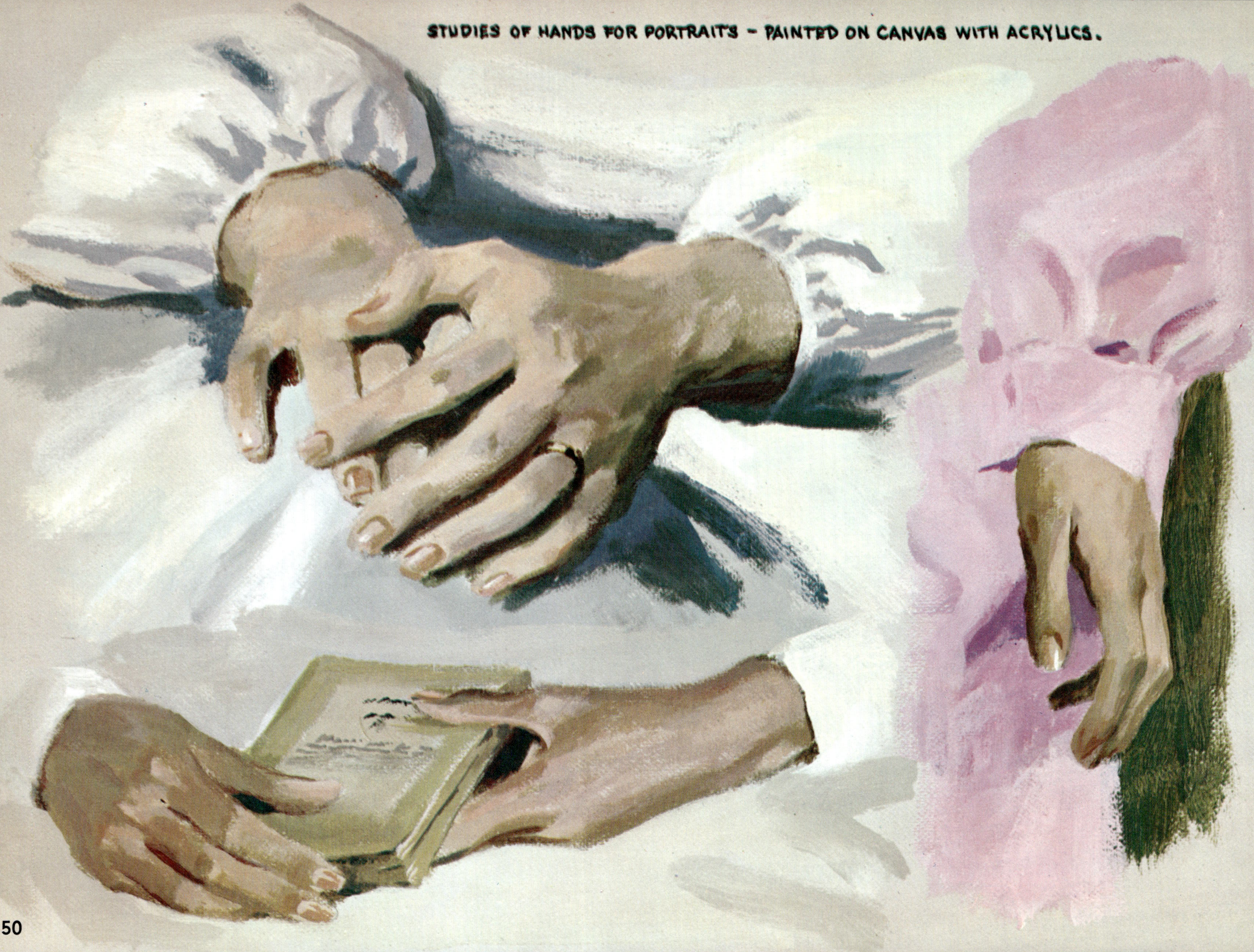

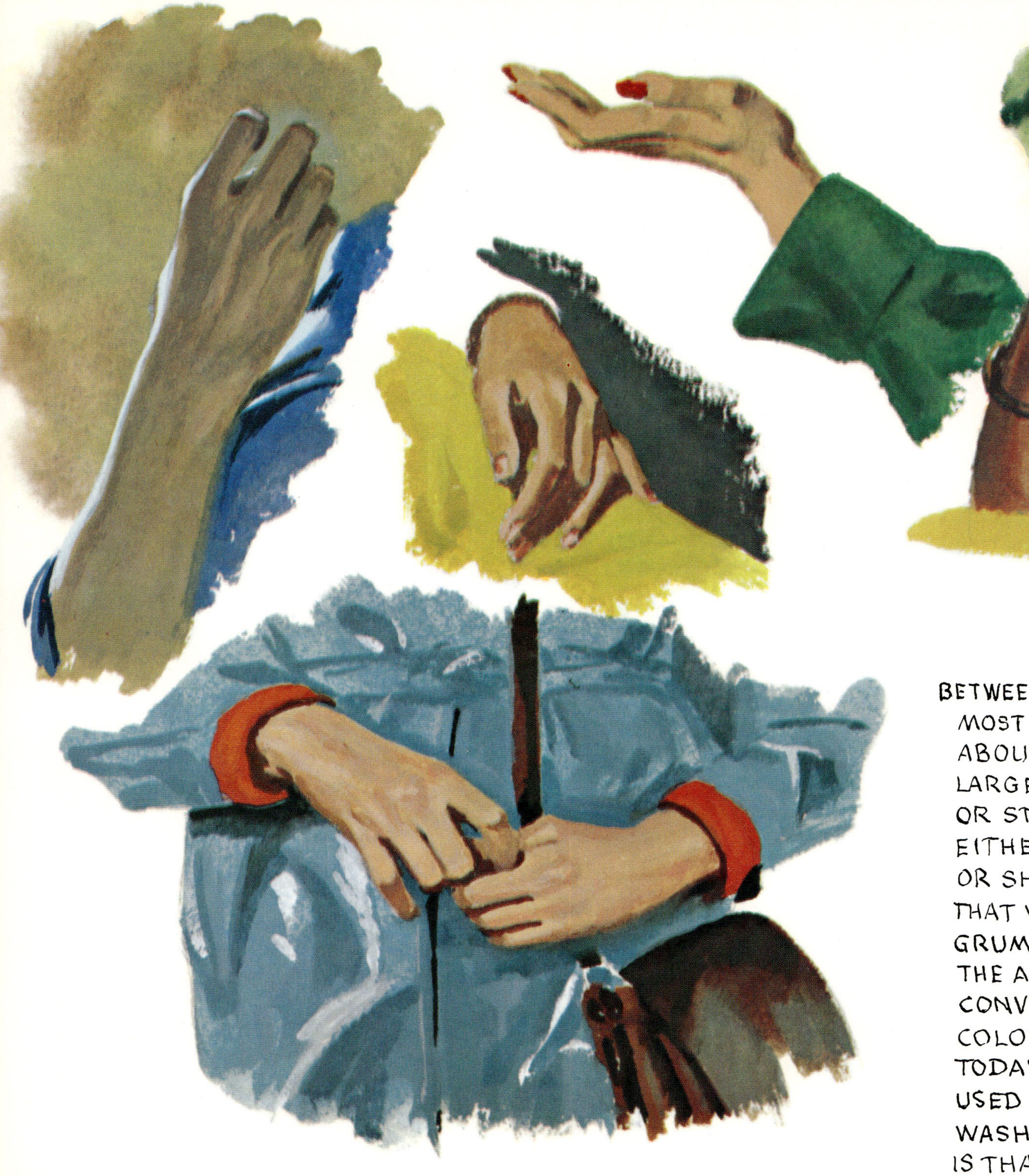

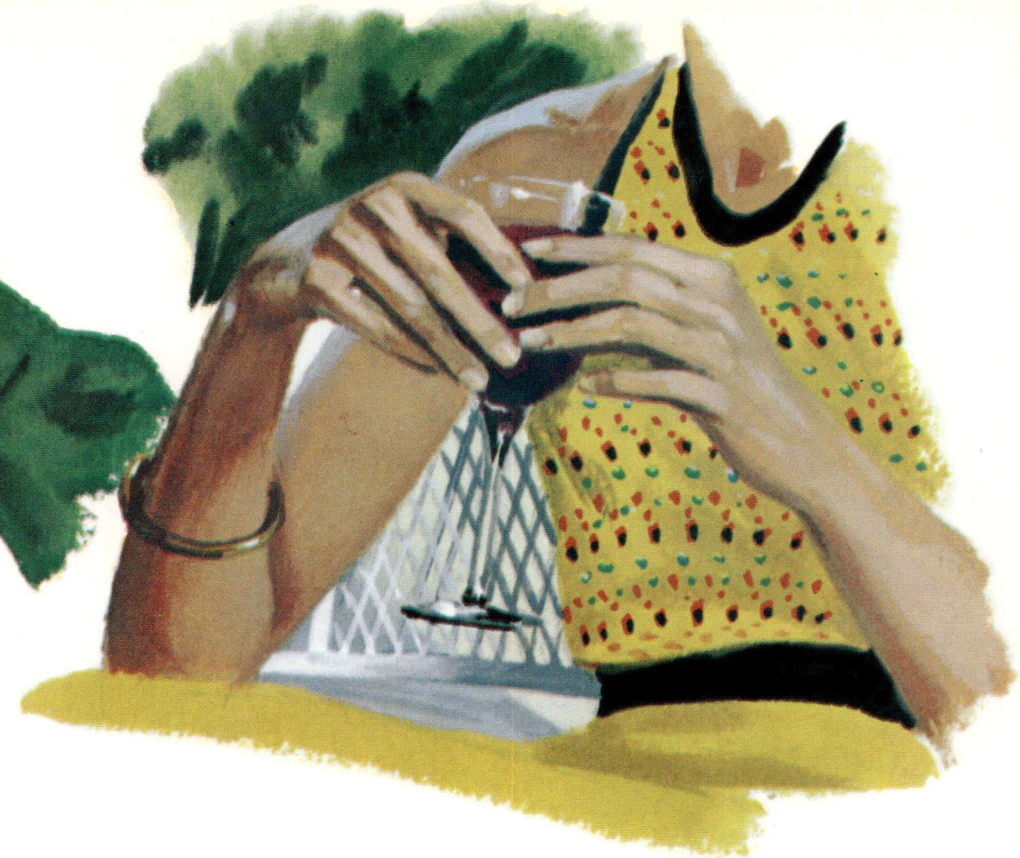

PAINTED ON MOUNTED WHATMAN COLD PRESSED WATER-COLOR PAPER WITH WINSOR & NEWTON "DESIGNERS'" COLORS.

BETWEEN THE YEARS OF THE LATE 1930'S TO THE MID 1950'S MOST OF US, WHO WERE ILLUSTRATORS, WORKED AT ABOUT THIS SCALE - SOME WORKED SLIGHTLY LARGER. MOST OF US PAINTED ON EITHER WHATMAN OR STRATHMORE ILLUSTRATION BOARD. WE USED EITHER WINSOR & NEWTON "DESIGNERS'" COLORS OR SHIVA CASEIN. THERE WERE OTHER BRANDS THAT WERE EQUALLY USED: ROWNEY, TALENS, GRUMBACHER AND OTHERS. WITH THE ADVENT OF THE ACRYLICS ON THE MARKET MANY ILLUSTRATORS CONVERTED TO THEIR USE. ACRYLICS ARE THE COLORS IN COMMON USE BY ILLUSTRATORS TODAY, THEIR ADVANTAGE IS THAT THEY CAN BE USED AS TRANSPARENTLY AS WATER-COLOR WASH OR AS THICKLY AS OILS. THEIR DISADVANTAGE IS THAT ONCE THEY HAVE DRIED THEY CAN NOT BE WASHED OUT.

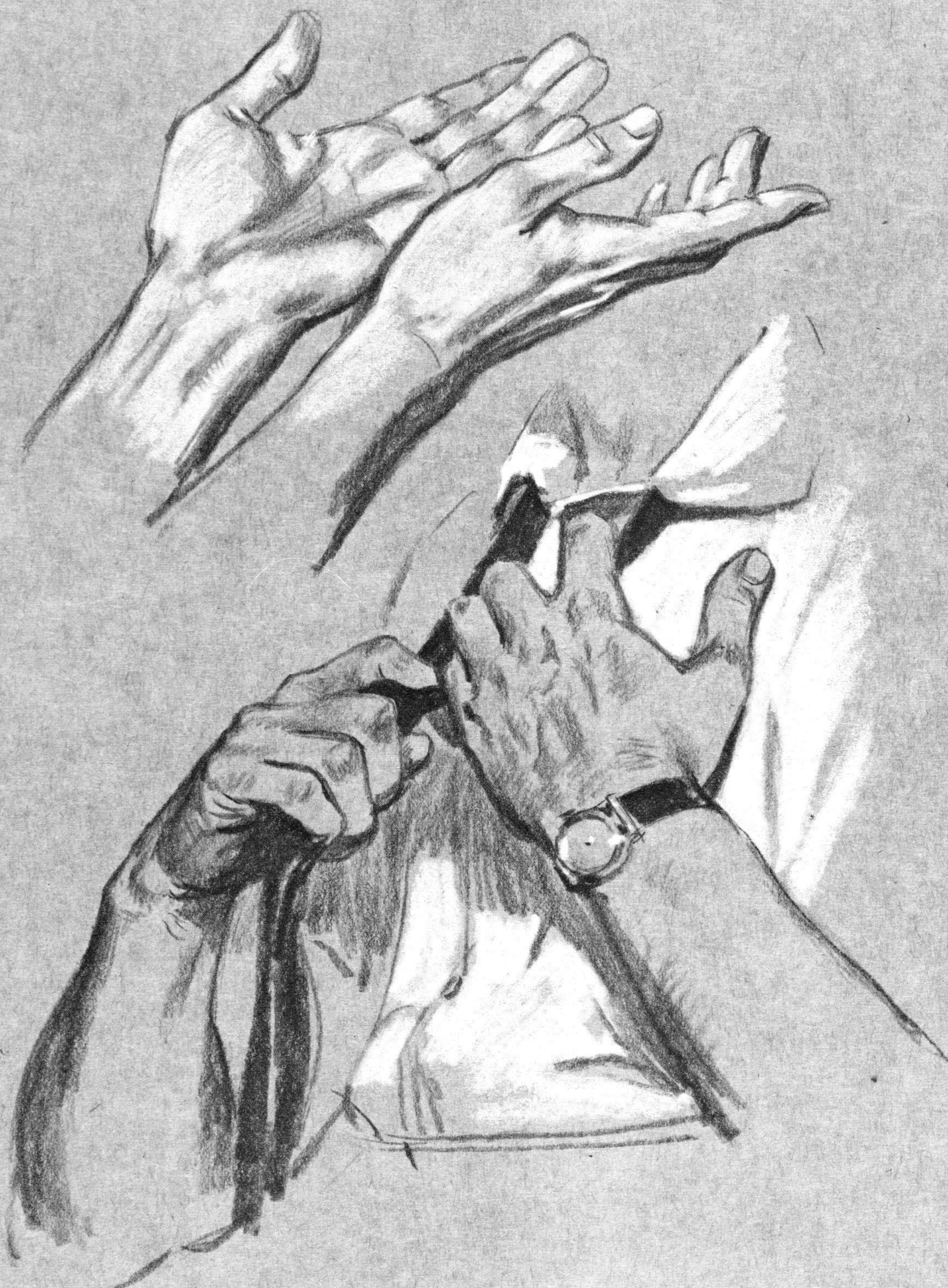

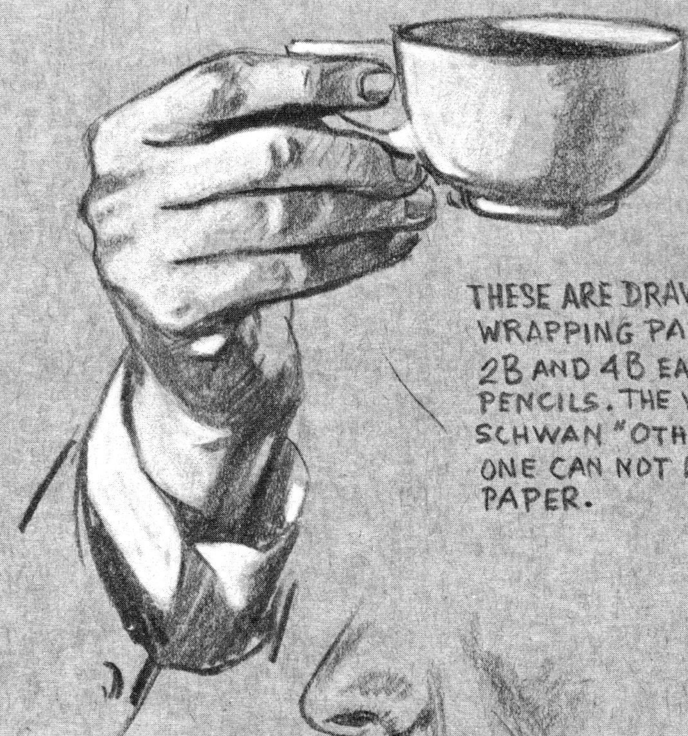

THESE ARE DRAWN ON HEAVY WRAPPING PAPER WITH HB, 2B AND 4B EAGLE "CHARCO" PENCILS. THE WHITE IS SCHWAN "OTHELLO" PENCIL. ONE CAN NOT ERASE ON THIS PAPER.

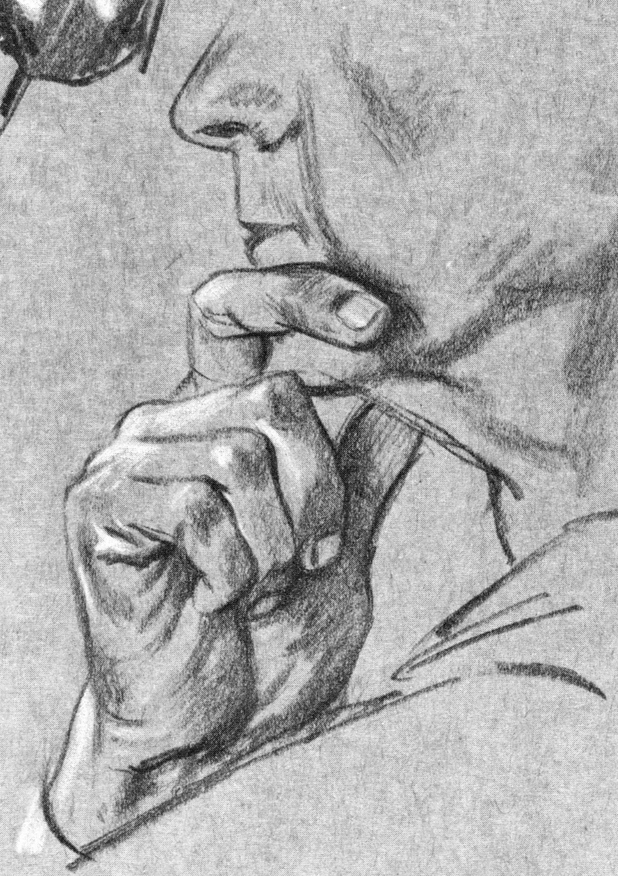

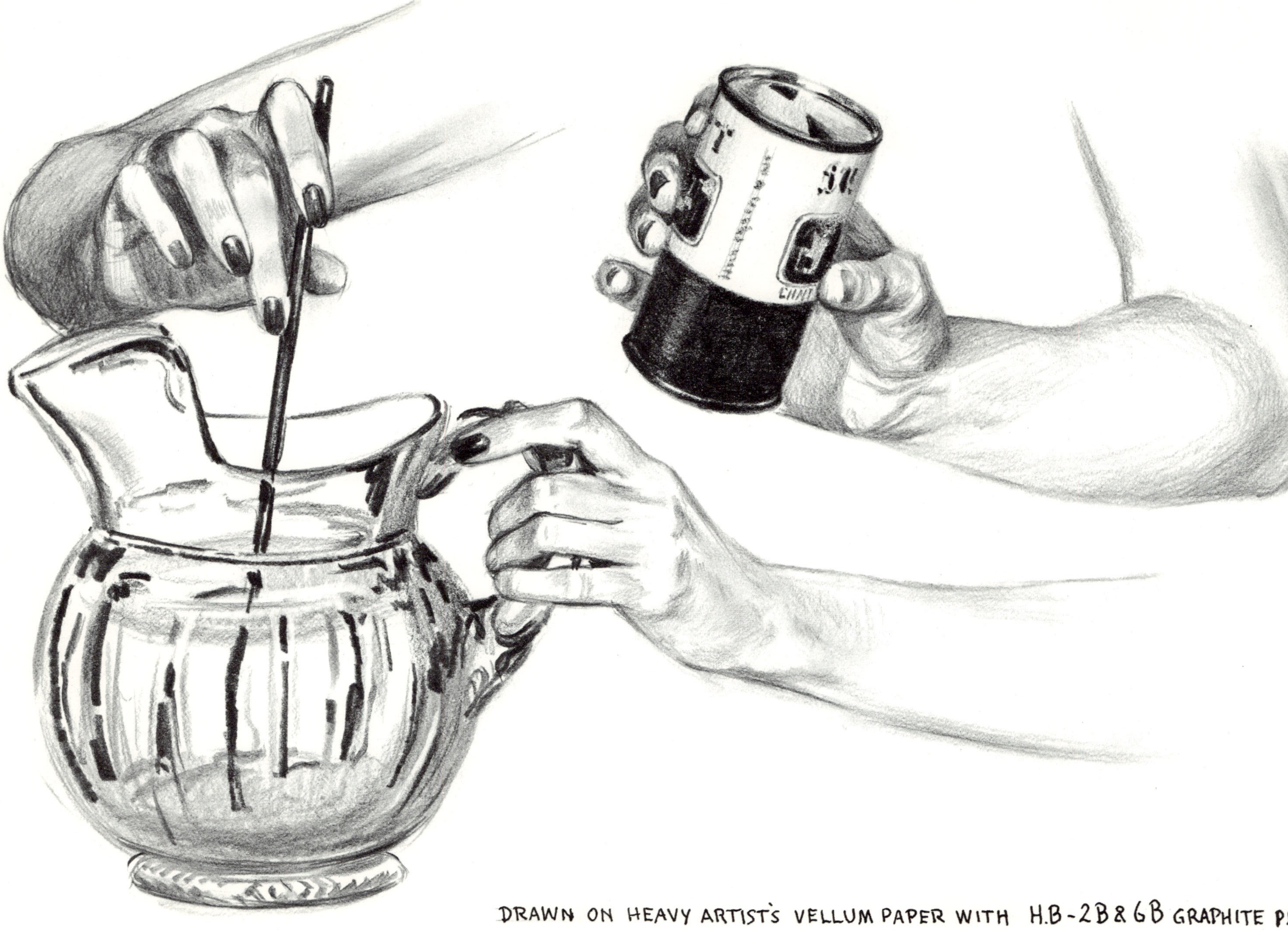

DRAWN ON HEAVY ARTIST'S VELLUM PAPER WITH H.B-2B & 6B GRAPHITE PENCIL.

DRAWN ON HEAVY ARTIST'S VELLUM WITH EAGLE HB, 2B AND 6B "CHARCO" PENCILS AND PAPER STUMP.

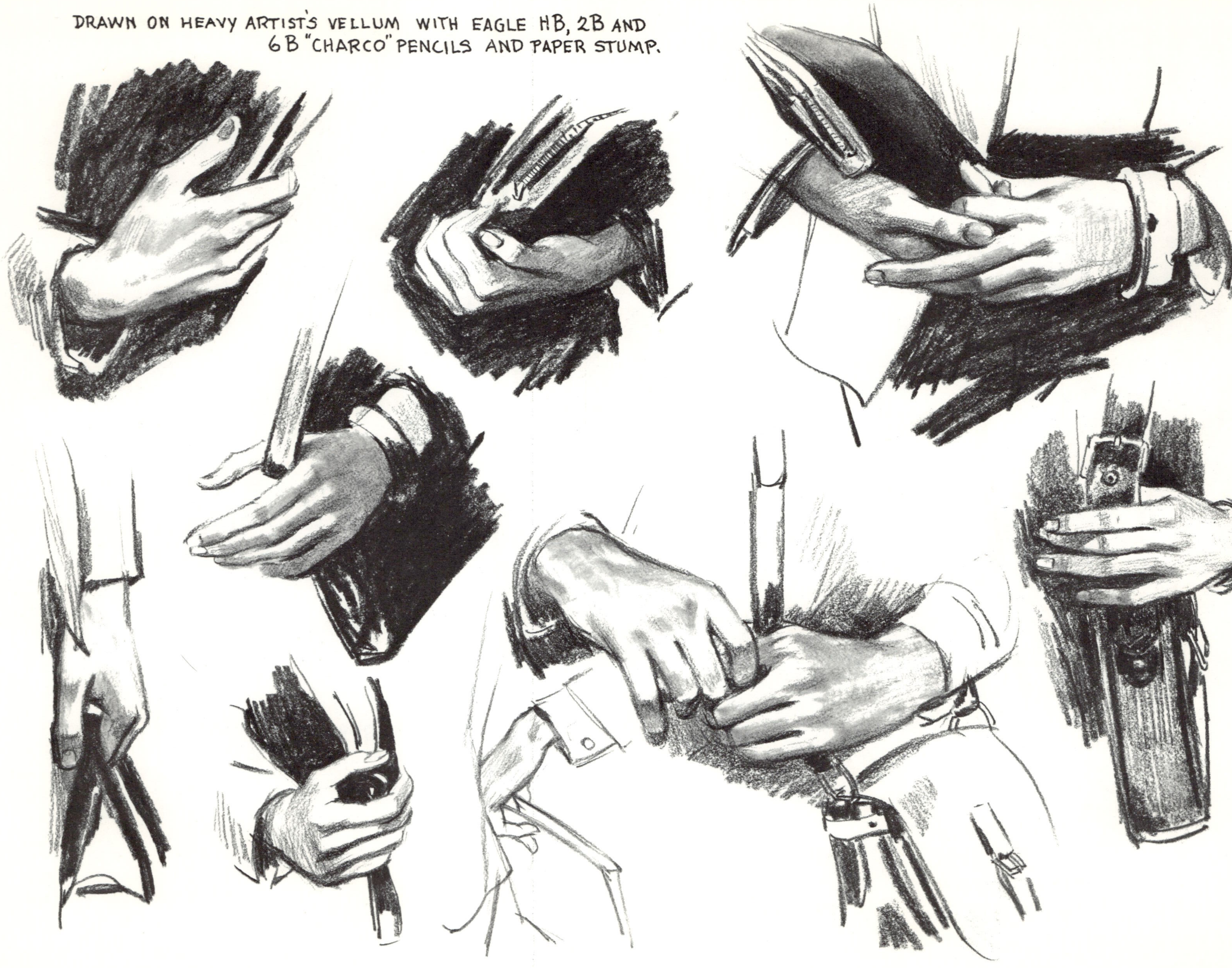

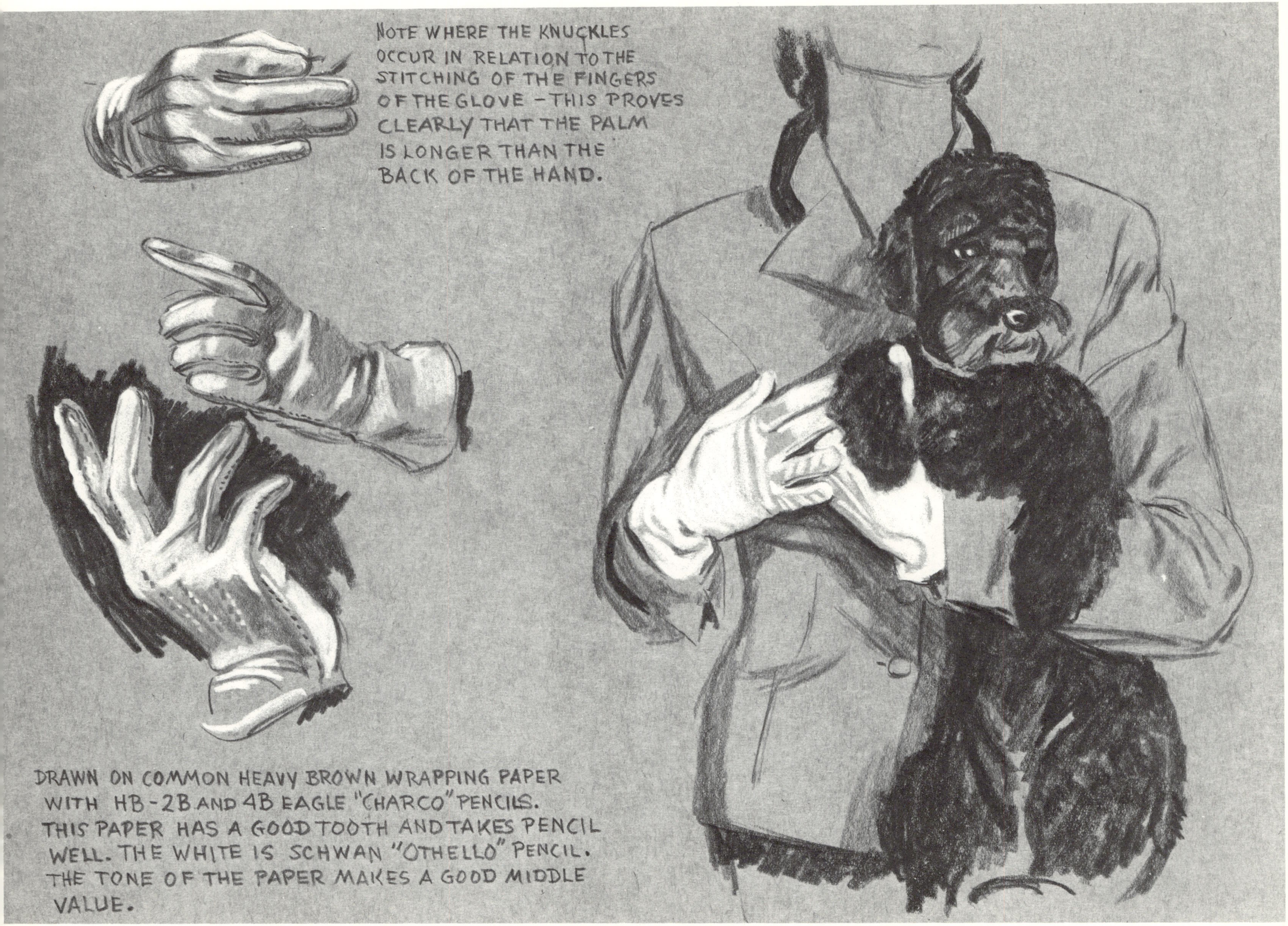
NOTE WHERE THE KNUCKLES OCCUR IN RELATION TO THE STITCHING OF THE FINGERS OF THE GLOVE - THIS PROVES CLEARLY THAT THE PALM IS LONGER THAN THE BACK OF THE HAND.
DRAWN ON COMMON HEAVY BROWN WRAPPING PAPER WITH HB-2B AND 4B EAGLE "CHARCO" PENCILS. THIS PAPER HAS A GOOD TOOTH AND TAKES PENCIL WELL. THE WHITE IS SCHWAN "OTHELLO" PENCIL. THE TONE OF THE PAPER MAKES A GOOD MIDDLE VALUE.

QUICK SKETCHES FROM A MODEL AND SOME FROM PHOTOS FROM NEWSPAPERS THIS IS A GOOD PRACTICE TO DEVELOP. FILL SKETCH-BOOKS WITH DRAWINGS OF THIS SCALE FOR QUICK GESTURE REFERENCE. MANY ILLUSTRATORS PAINT AT THIS SCALE.

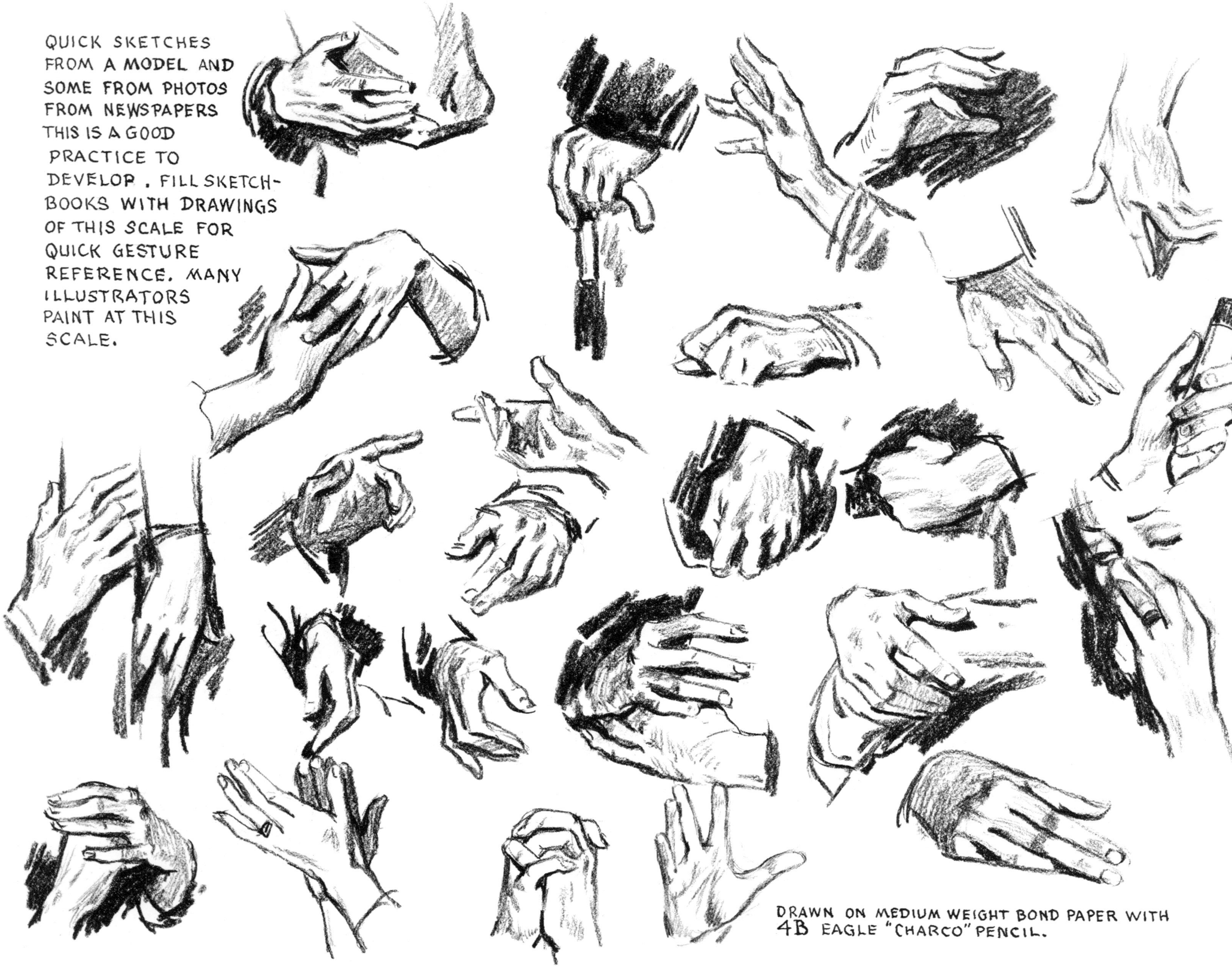

DRAWN ON MEDIUM WEIGHT BOND PAPER WITH 4B EAGLE "CHARCO" PENCIL.

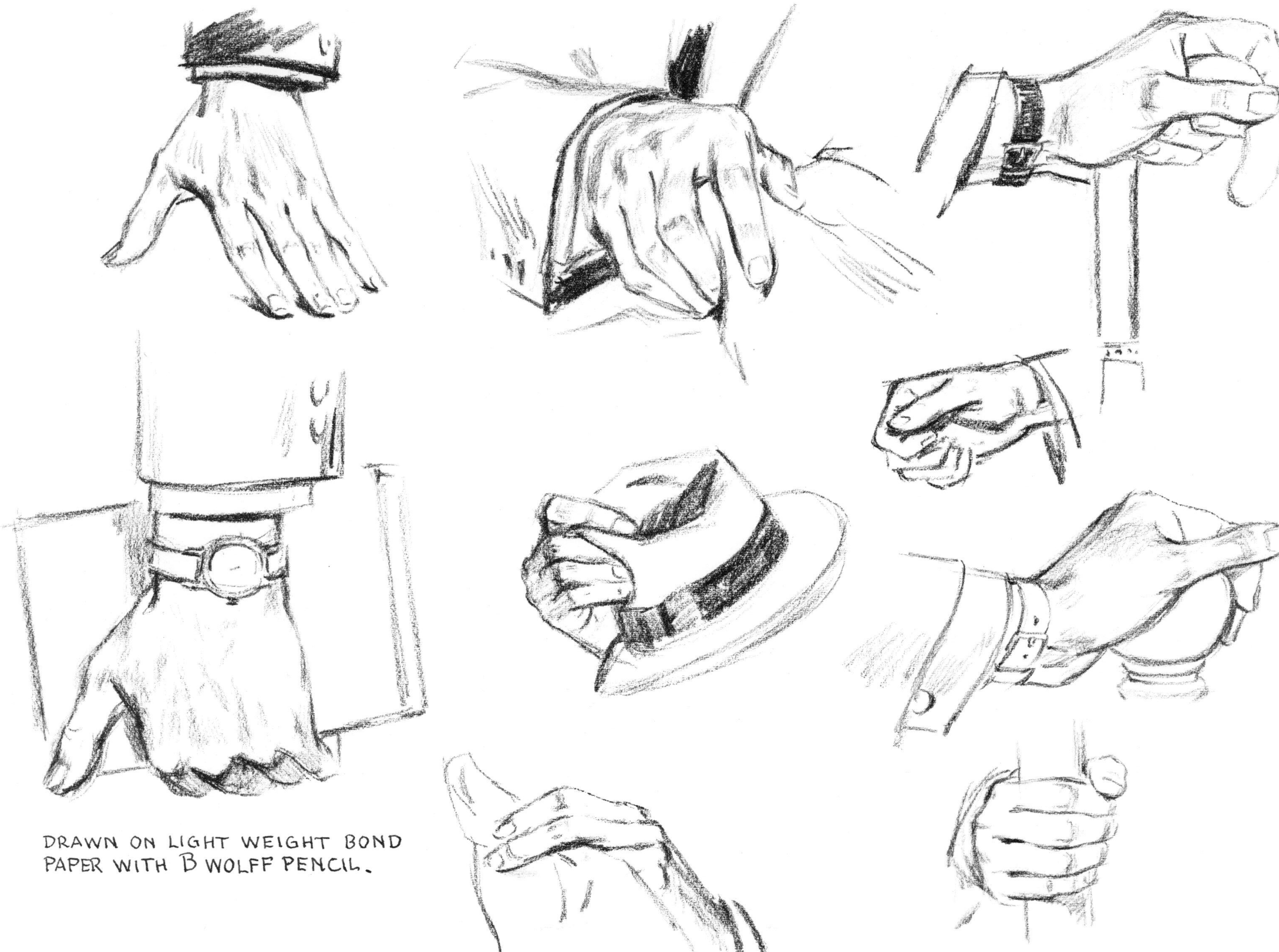

DRAWN ON LIGHT WEIGHT BOND PAPER WITH B WOLFF PENCIL.

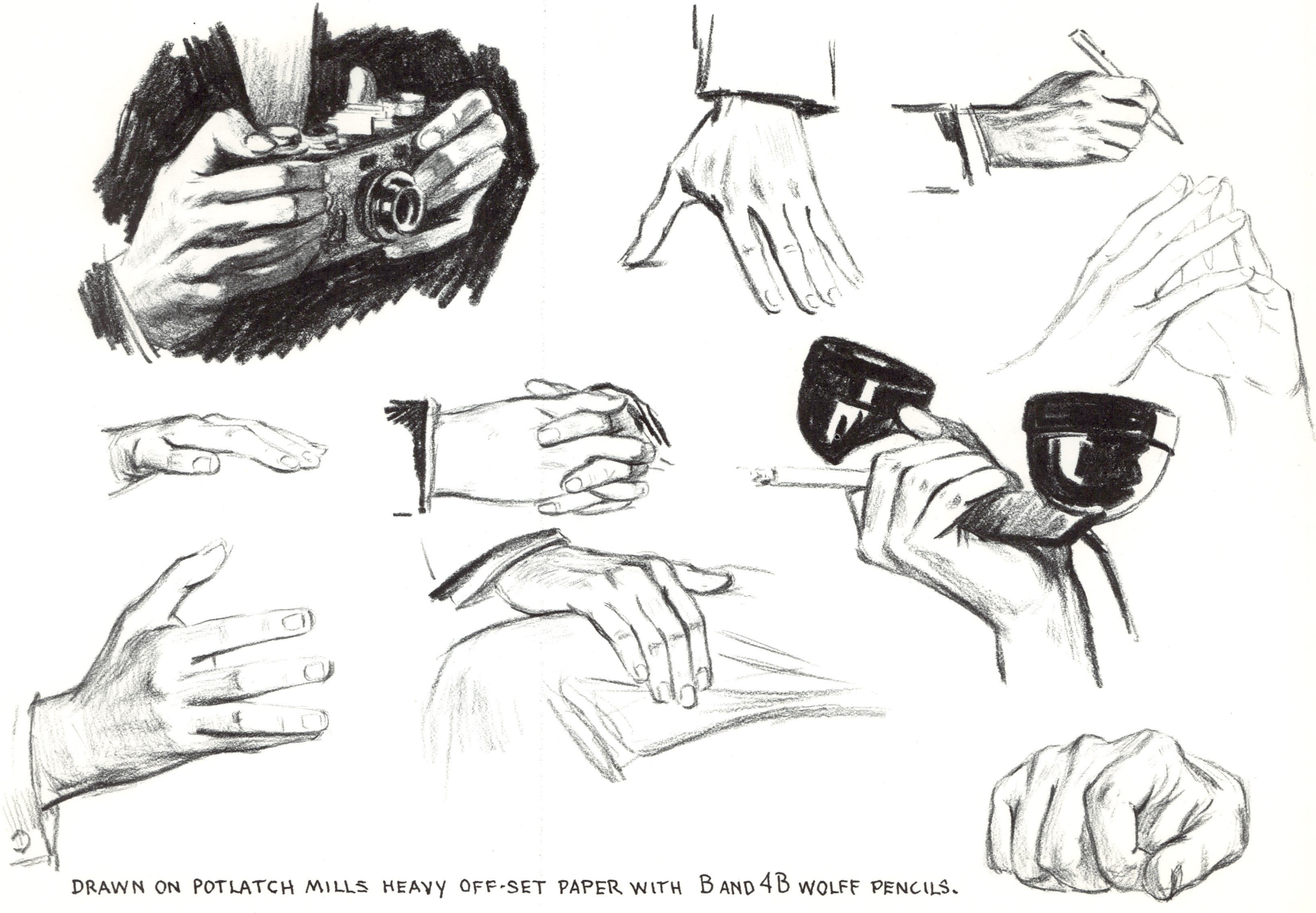

DRAWN ON POTLATCH MILLS HEAVY OFF-SET PAPER WITH B AND 4B WOLFF PENCILS.

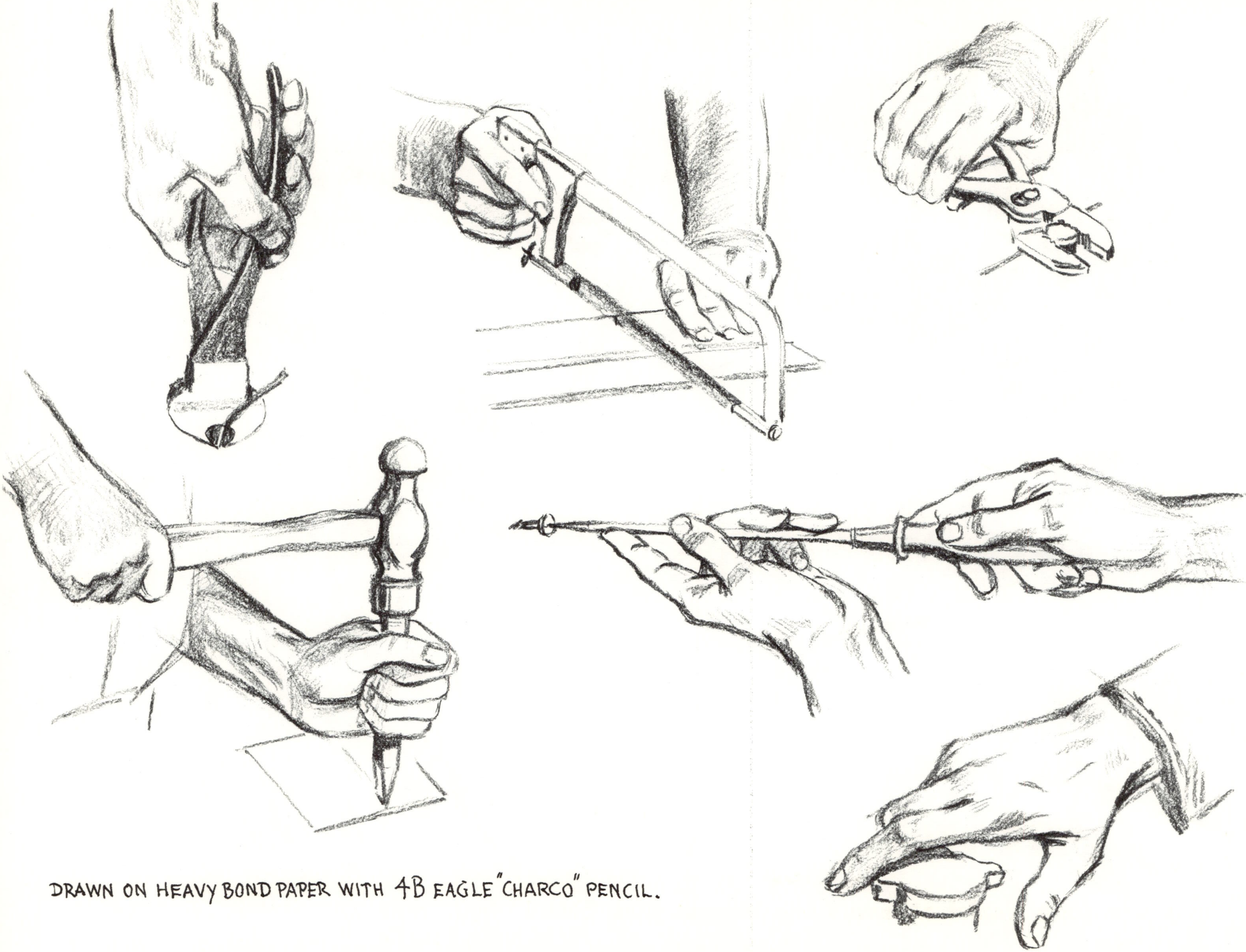

DRAWN ON HEAVY BOND PAPER WITH 4B EAGLE "CHARCO" PENCIL.

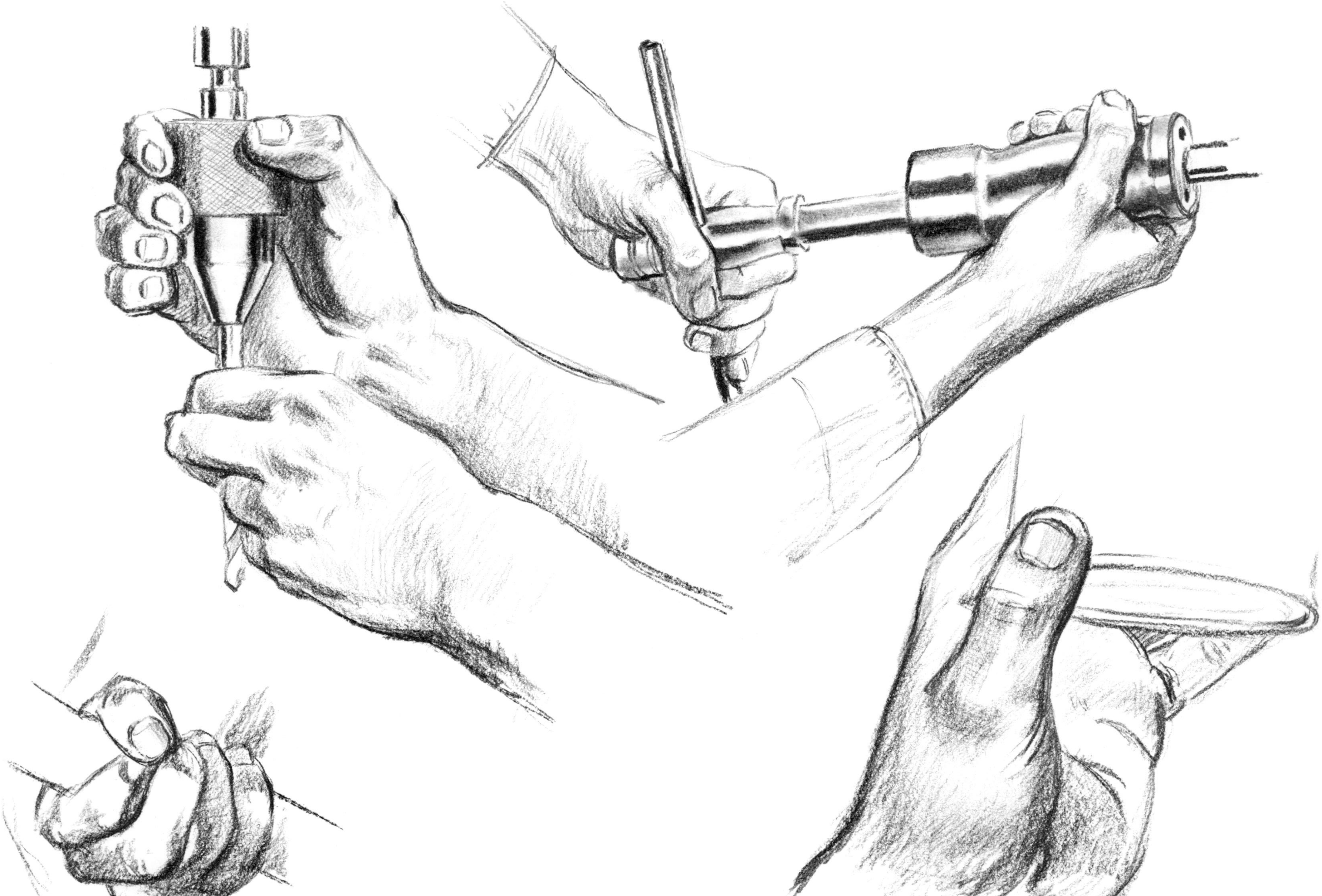

DRAWN ON HEAVY BOND PAPER WITH HB& 4B EAGLE "CHARCO" PENCIL.

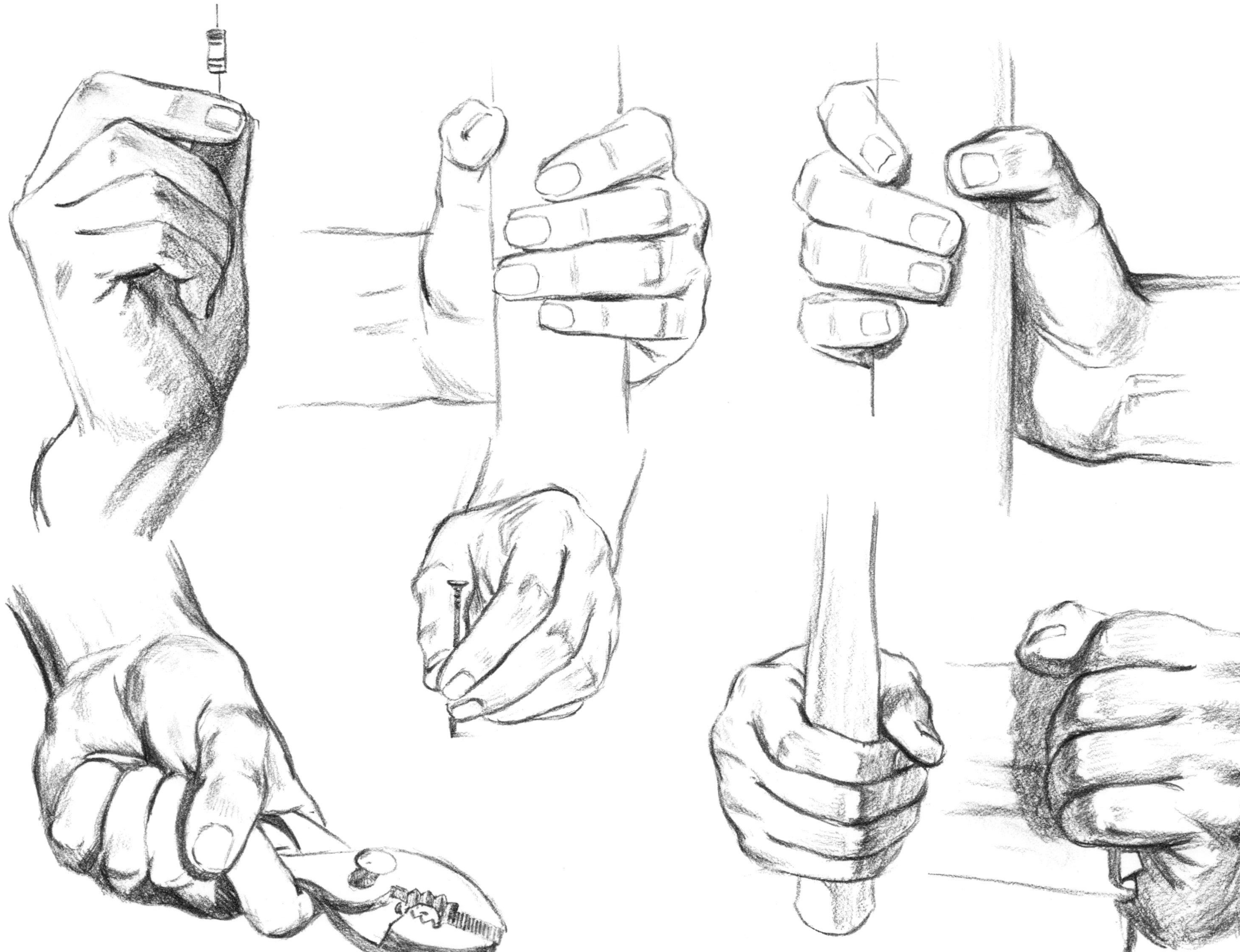

DRAWN ON HEAVY ARTIST'S VELLUM WITH 2B WOLFF PENCIL.

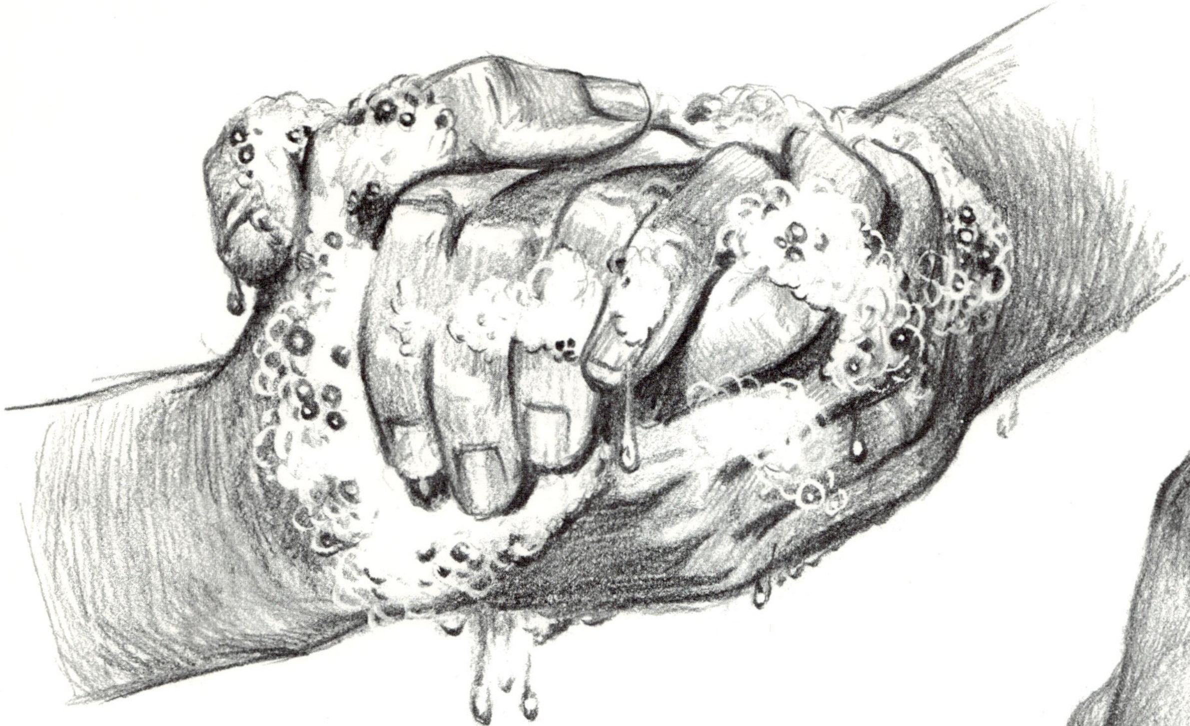

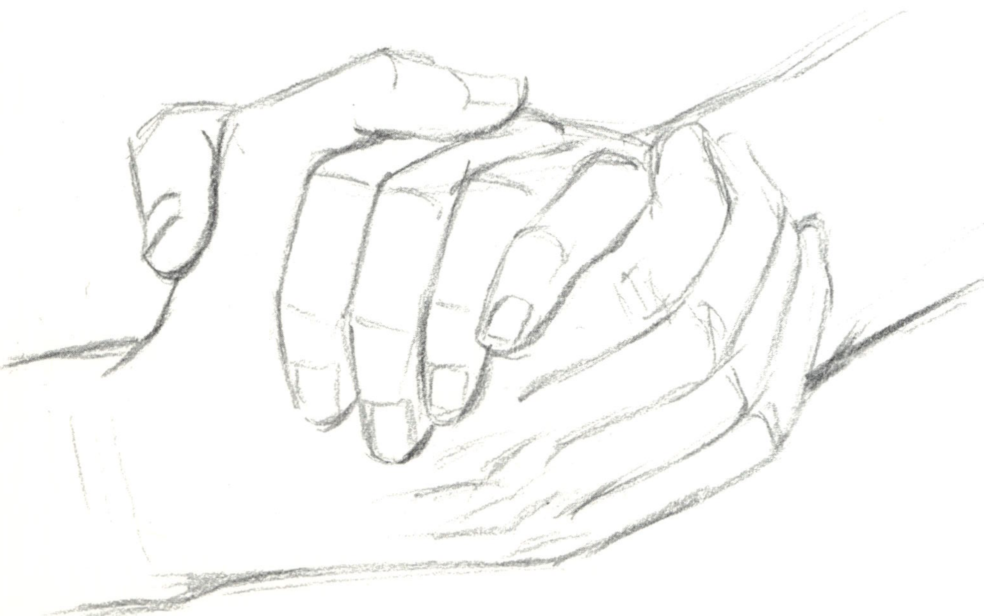

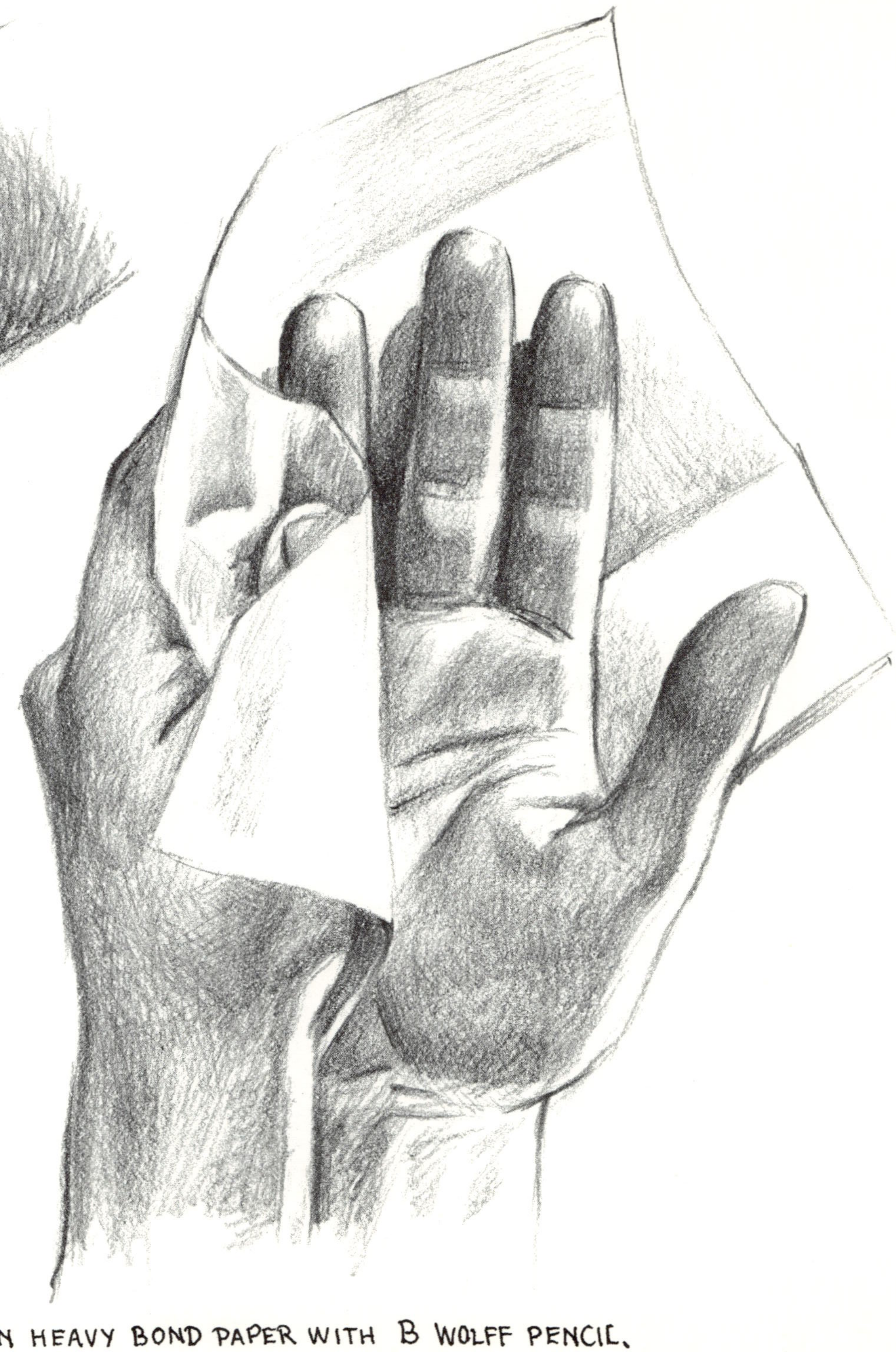

DRAWN ON HEAVY BOND PAPER WITH B WOLFF PENCIL.

DRAWN ON STRATHMORE CHARCOAL PAPER WITH 2B & 6B GRAPHITE PENCILS.

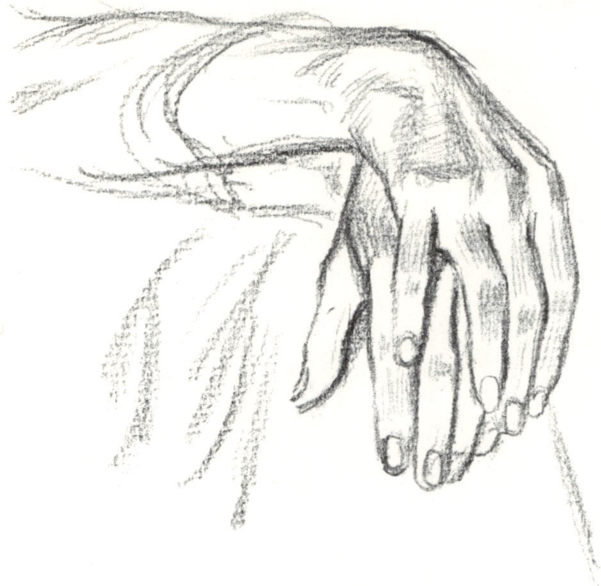

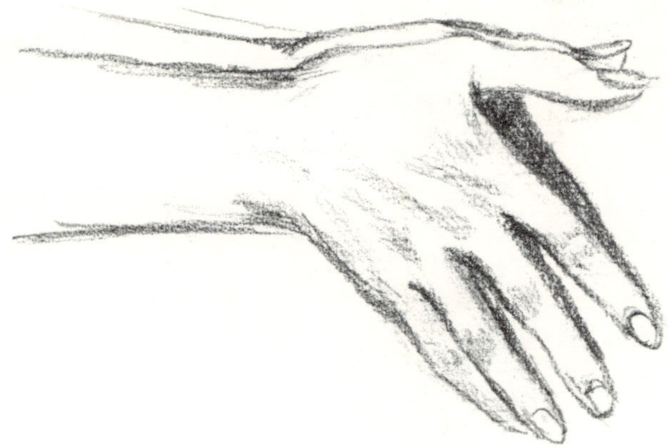

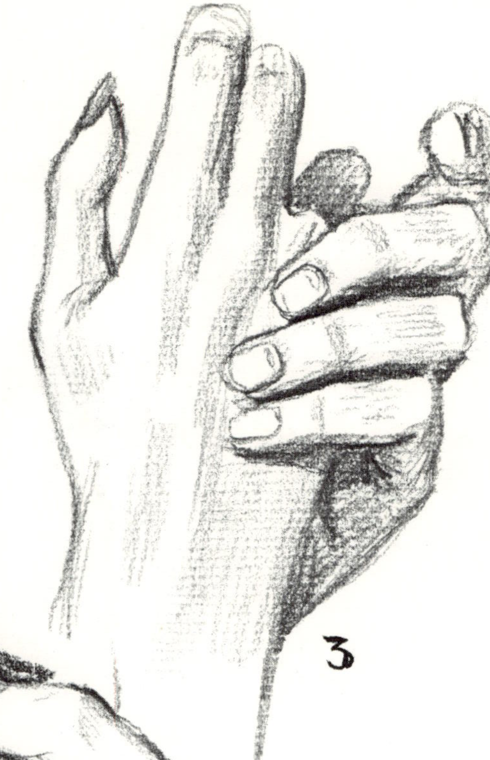

3

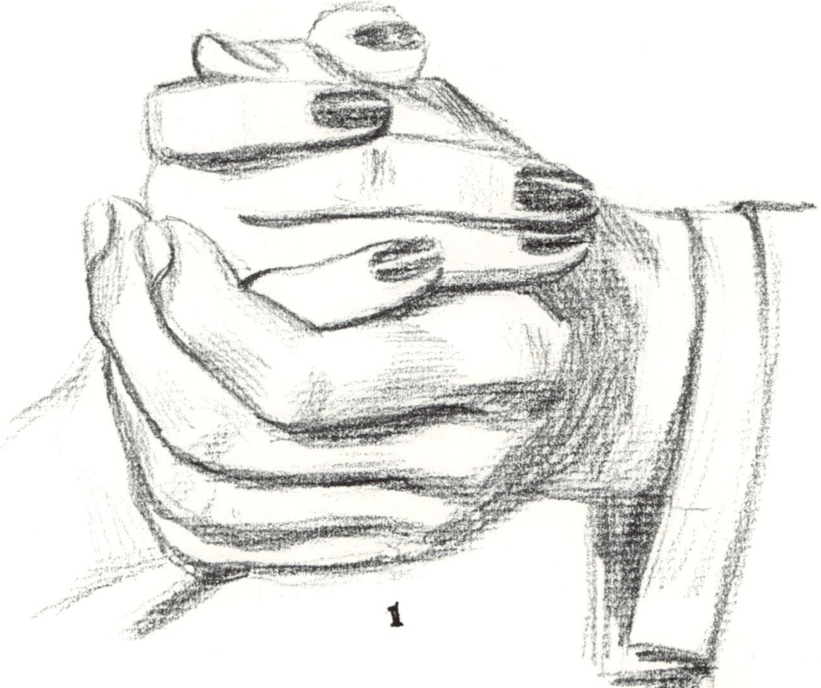

1

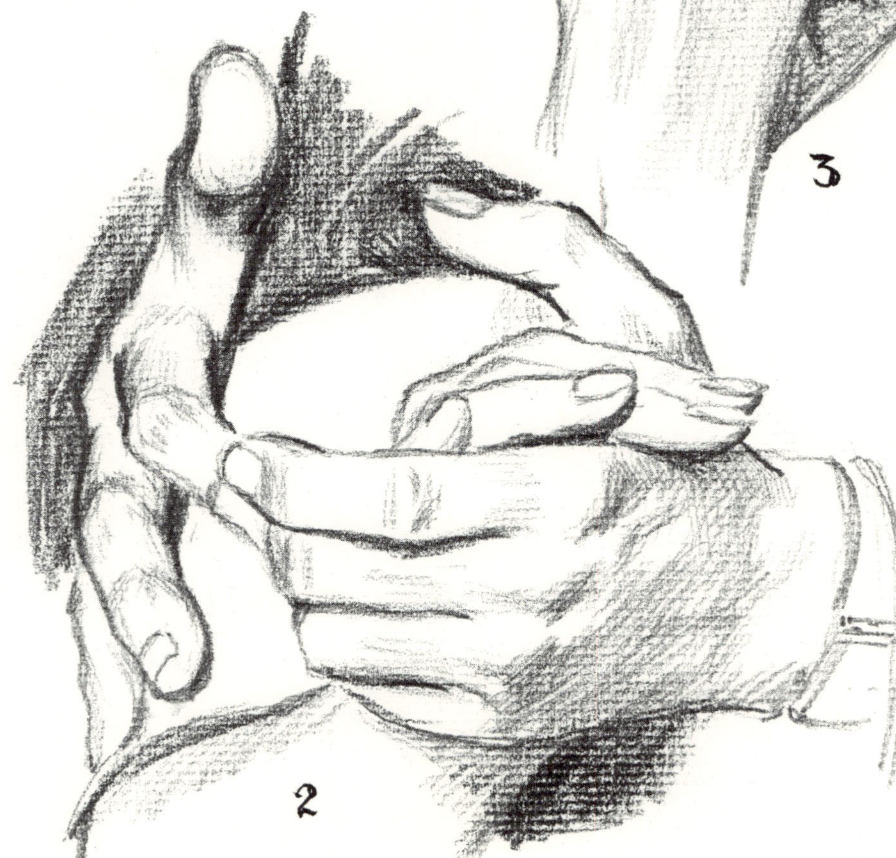

2

DRAWINGS 1, 2 & 3 ARE HAND GESTURES OF BALLROOM DANCING THAT HAS NOW BECOME, IT SEEMS, AN ARCHAIC FORM OF SOCIAL AMUSEMENT. THE HABITS CHANGE - AH ME!

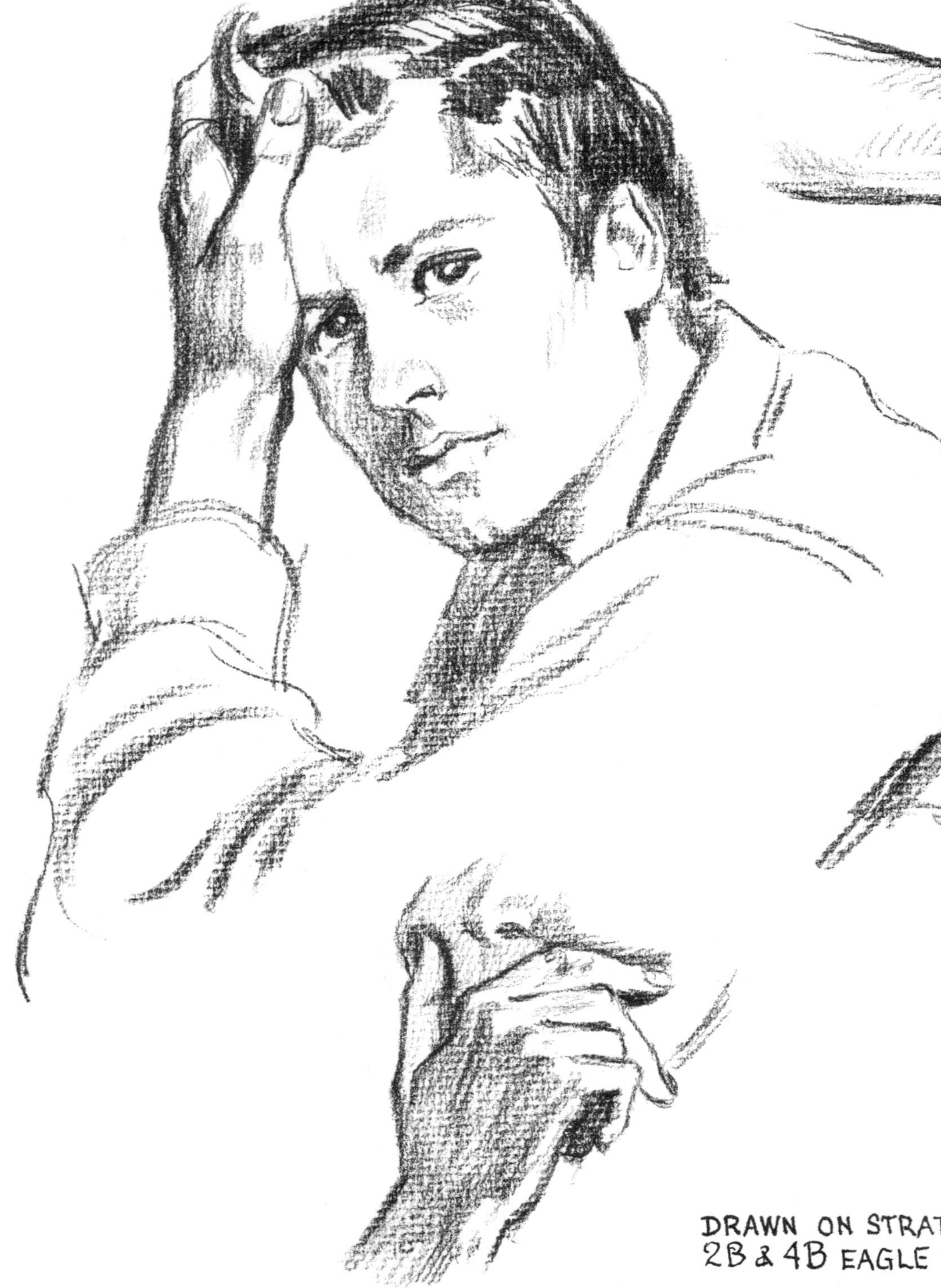

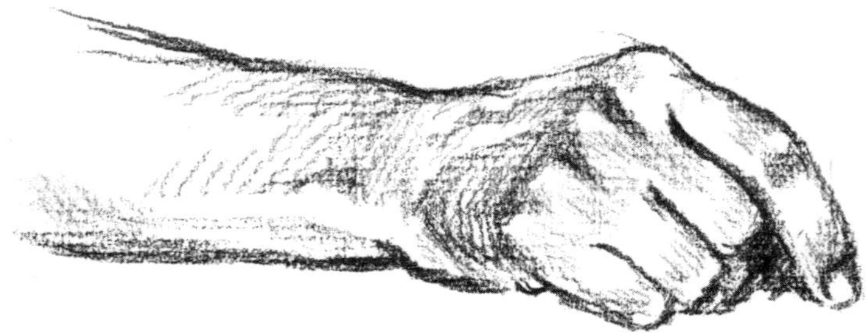

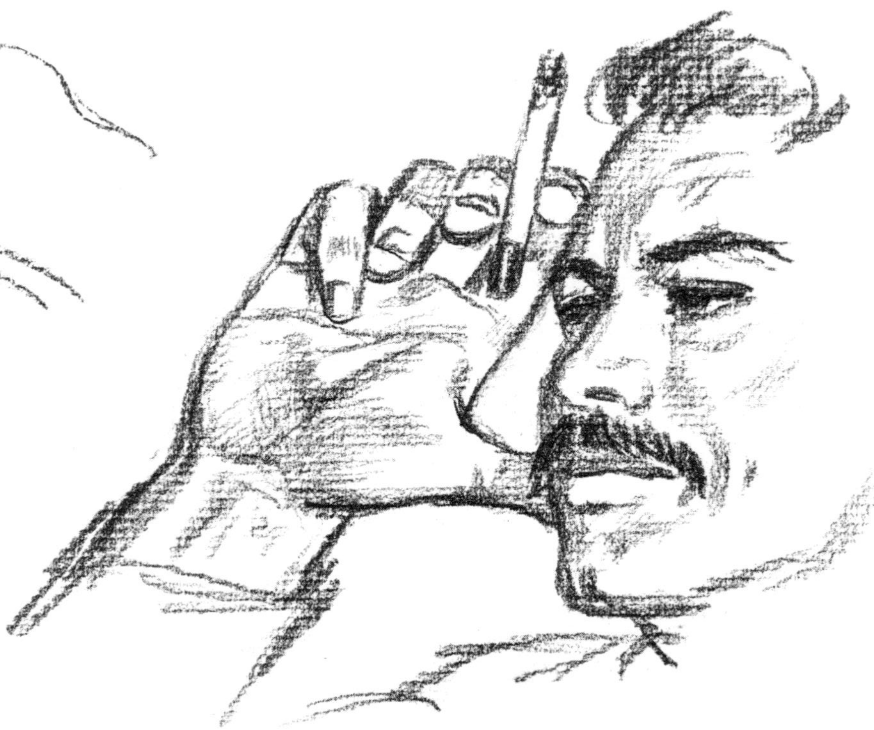

DRAWN ON STRATHMORE CHARCOAL PAPER WITH
2B & 4B EAGLE "CHARCO" PENCILS.

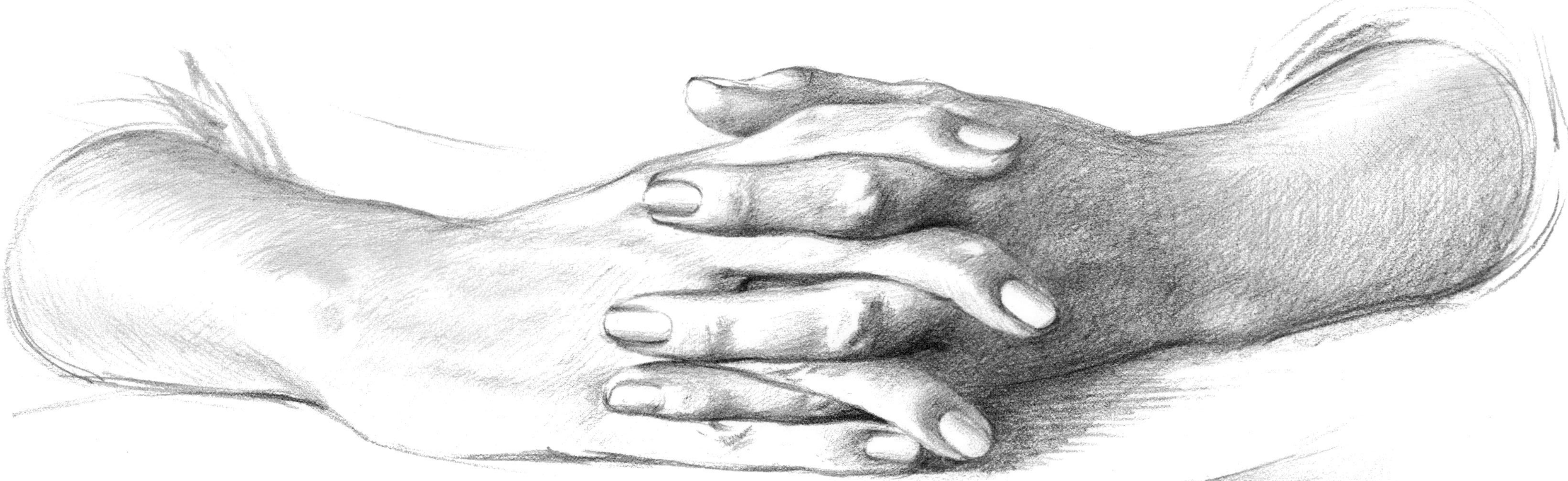

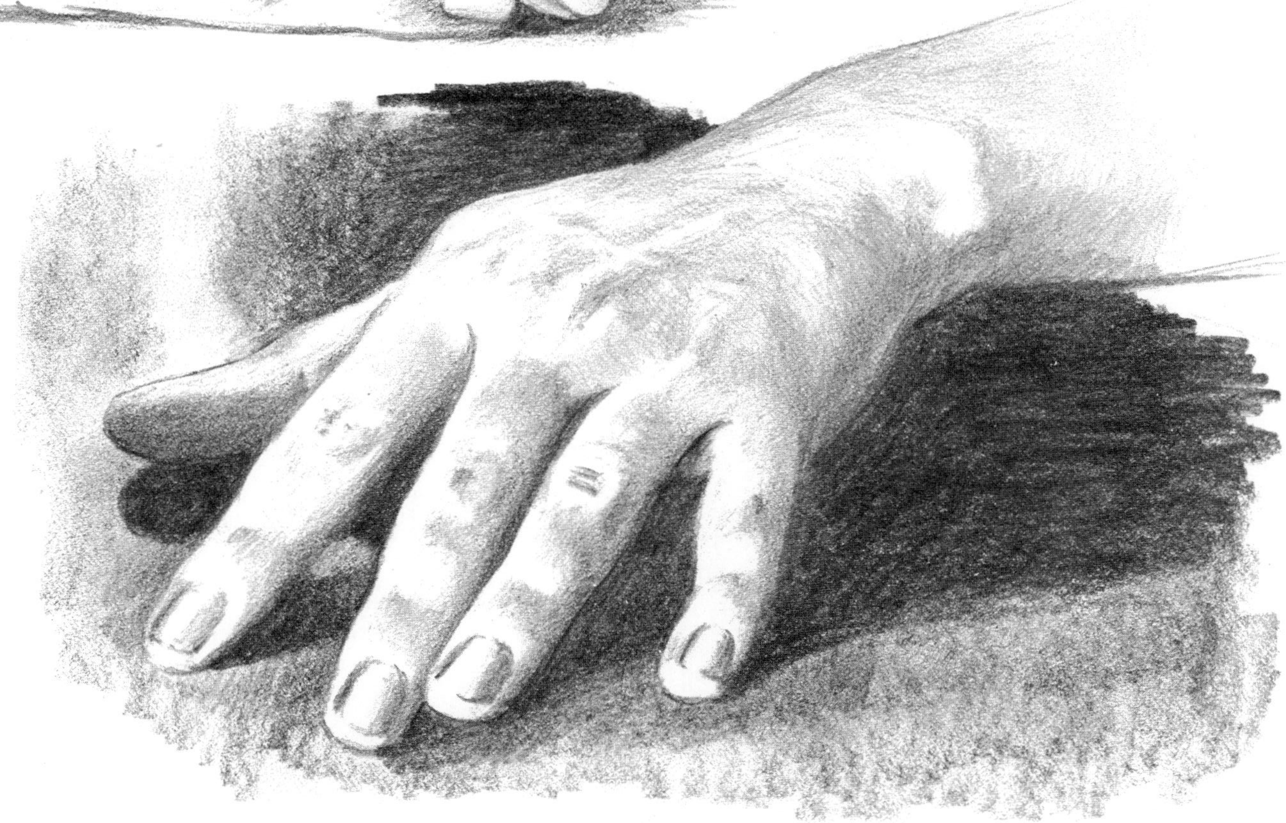

Full tonal value studies of this type are often made as reference material when one is painting a portrait. I am sure that photos of the sitter's hands might serve just as well — however, for me, it seems that studies made from the sitter have more character. One seems to edit the points which impress one most by drawing from the sitter instead of using photos.

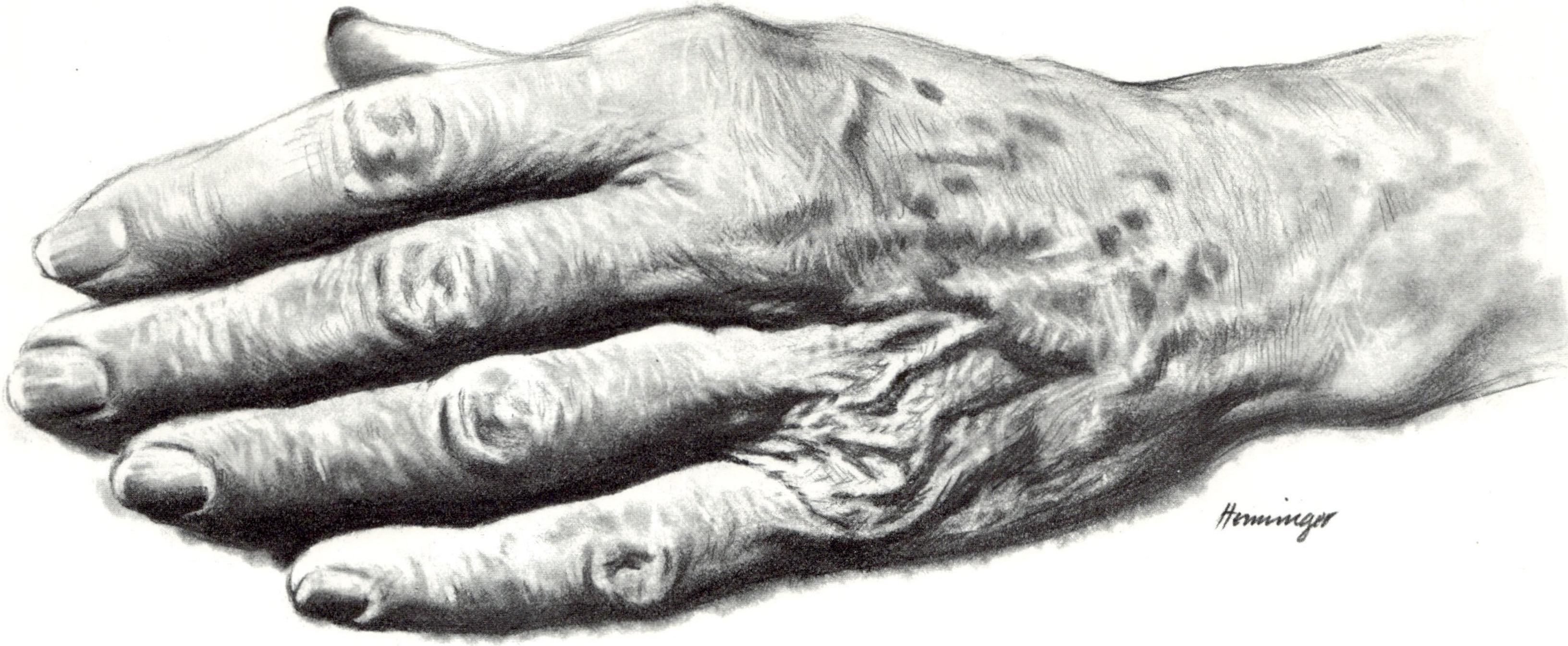

THIS MODE OF DRAWING IS A GOOD, OCCASIONAL, EXERCISE BUT IS NOT RECOMMENDED AS A REGULAR PRACTICE. IT IS INTERESTING TO SEE HOW WELL YOU CAN DISCIPLINE YOUR HAND AND EYE IN PRODUCING IMITATIVE VALUE (SHADING) AND TEXTURES. THE CAMERA IS THE BEST RECORDER OF DETAIL AND CAN DO IT FASTER. DRAWINGS OF THIS TYPE SHALL CHARM YOUR FAMILY AND FRIENDS WITH YOUR CLEVERNESS BUT WILL GET LITTLE, IF ANY, PRAISE FROM ART CRITICS.

DRAWN ON HEAVY BOND PAPER WITH "B" TO "6B" GRAPHITE PENCILS AND PAPER STUMP.

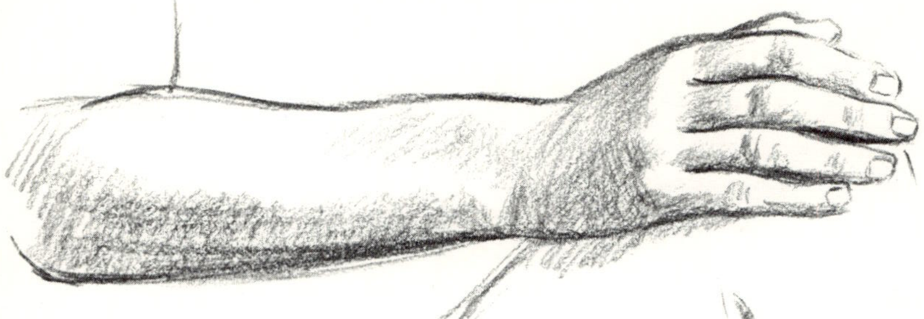

7 YEAR OLD BOY

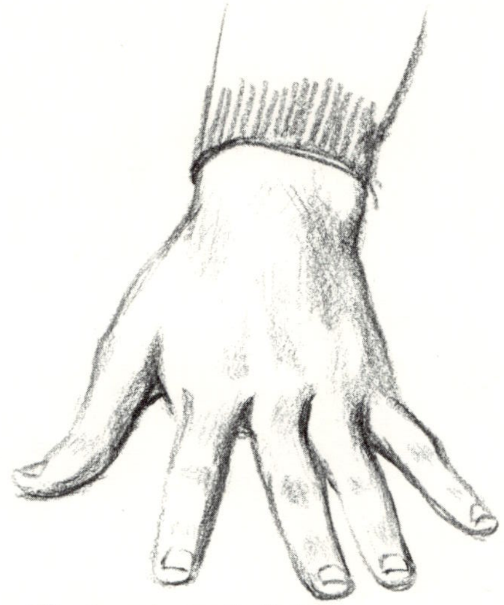

12 YEAR OLD BOY

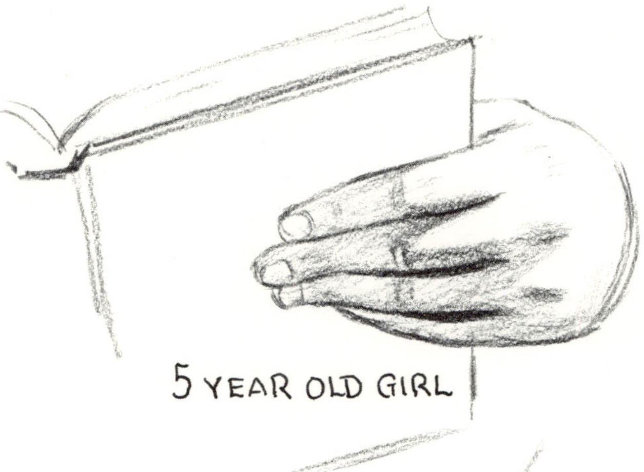

5 YEAR OLD GIRL

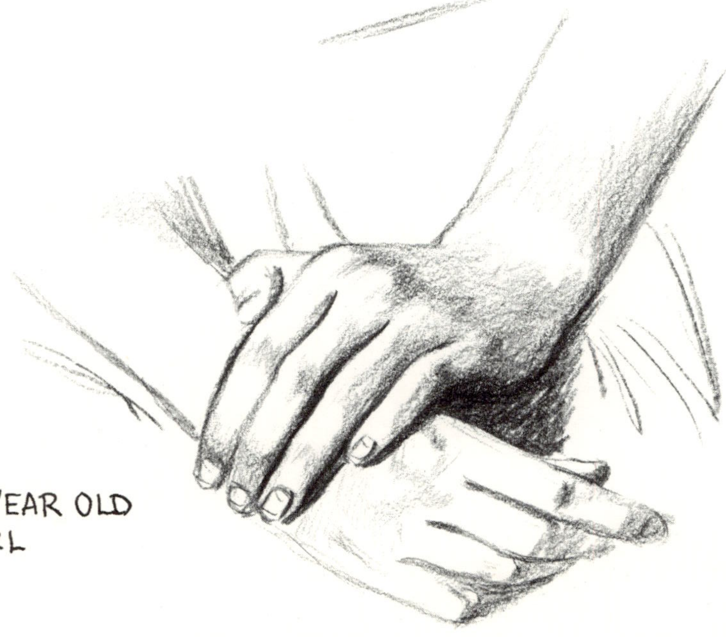

12 YEAR OLD GIRL

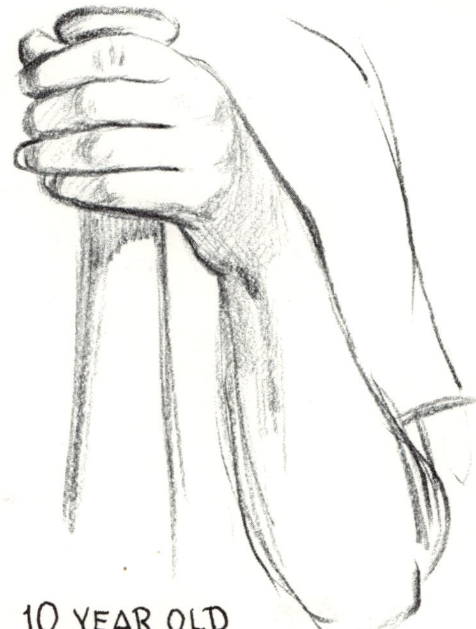

10 YEAR OLD BOY

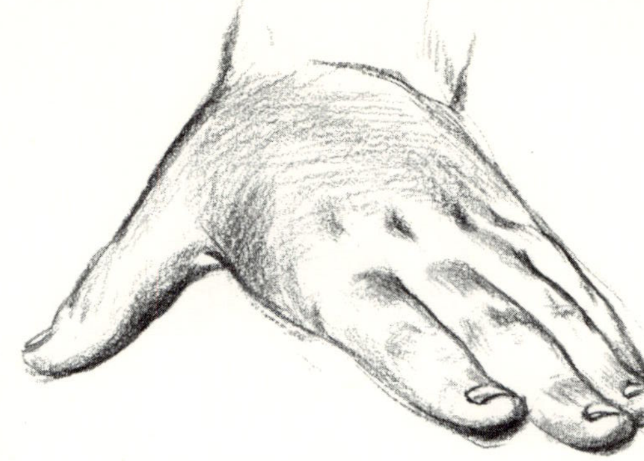

THE END PHALANGE OF A CHILD'S FINGERS IS QUITE FLEXIBLE AND HAS A TENDENCY TO BEND BACKWARDS. IN EARLY AGE A GIRL'S HAND AND BOY'S HAND CAN NOT BE TOLD APART.

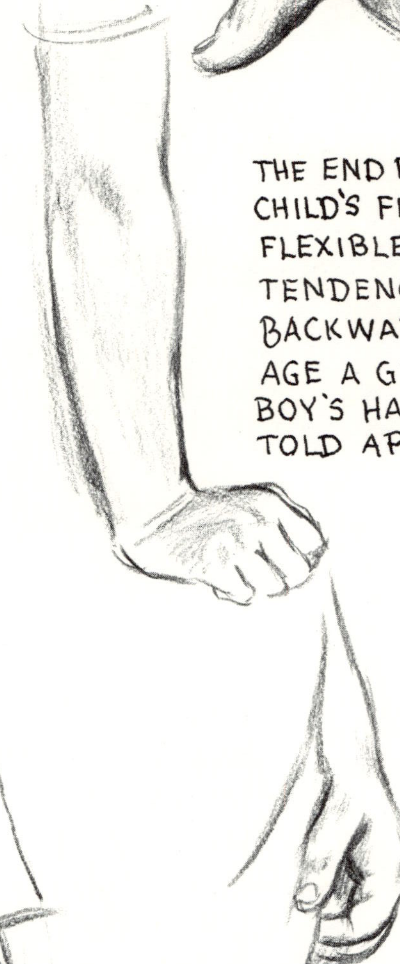

4 YEAR OLD GIRL

DRAWN ON HEAVY BOND PAPER WITH H.B. WOLFF PENCIL.

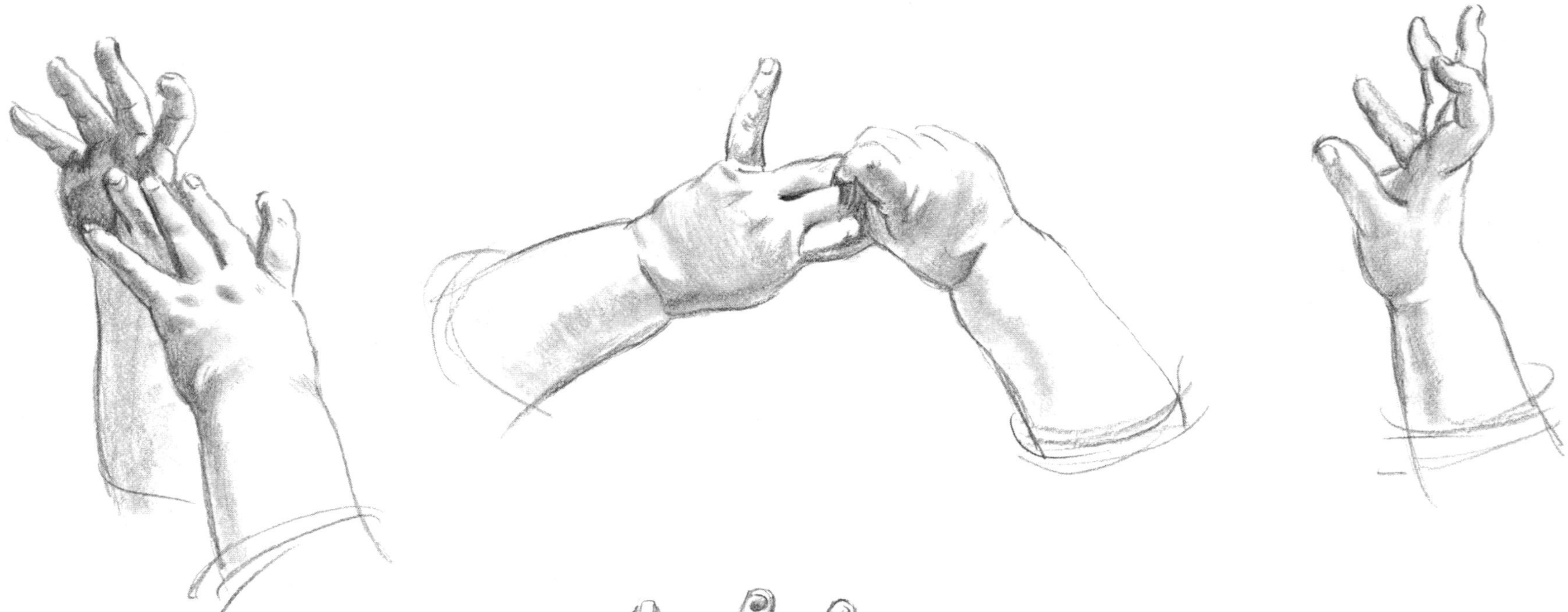

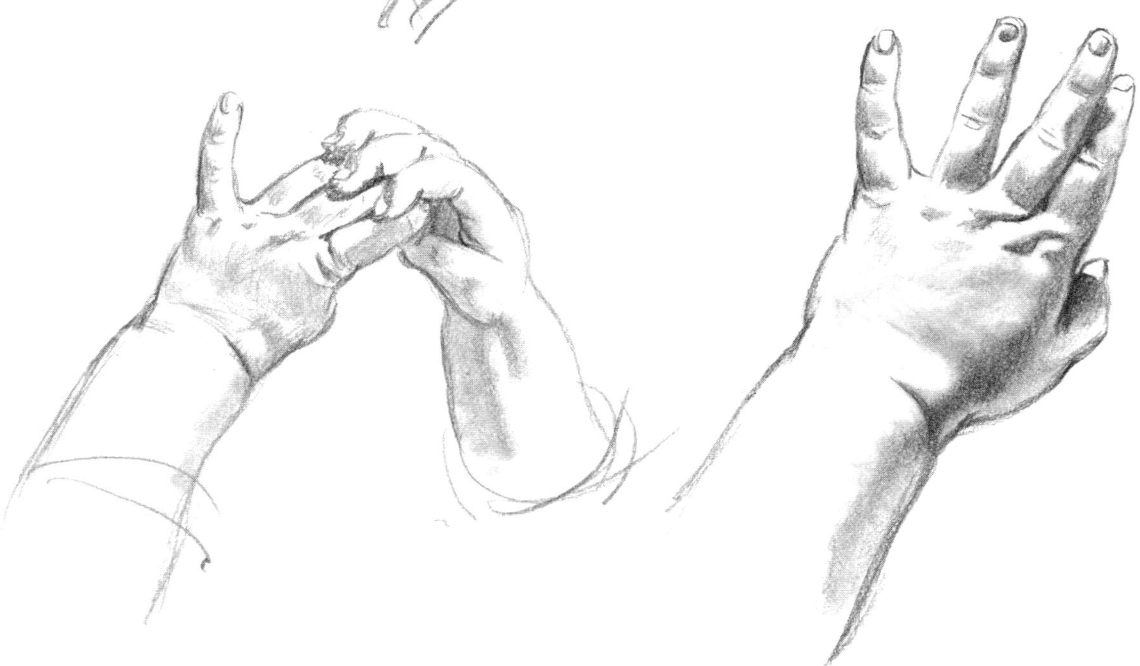

The basic shape of a baby's fingers can be characterized as conical; the tips are rounded and the fingernails are very small. The flexibility of a baby's hand and fingers is due to the fact that the bones are cartilaginous. Ossification starts between the first and second years. The wrist bones (carpal bones) do not begin to form until the second year and do not develop fully until the ninth year. Note, however, in the drawings on this page that the area that these bones shall ultimately occupy is clearly defined. The fatty tissue that covers the fingers and hand of a baby causes a deep dimpling at all the knuckles.

DRAWN ON HEAVY BOND PAPER WITH 2B GRAPHITE PENCIL AND PAPER STUMP

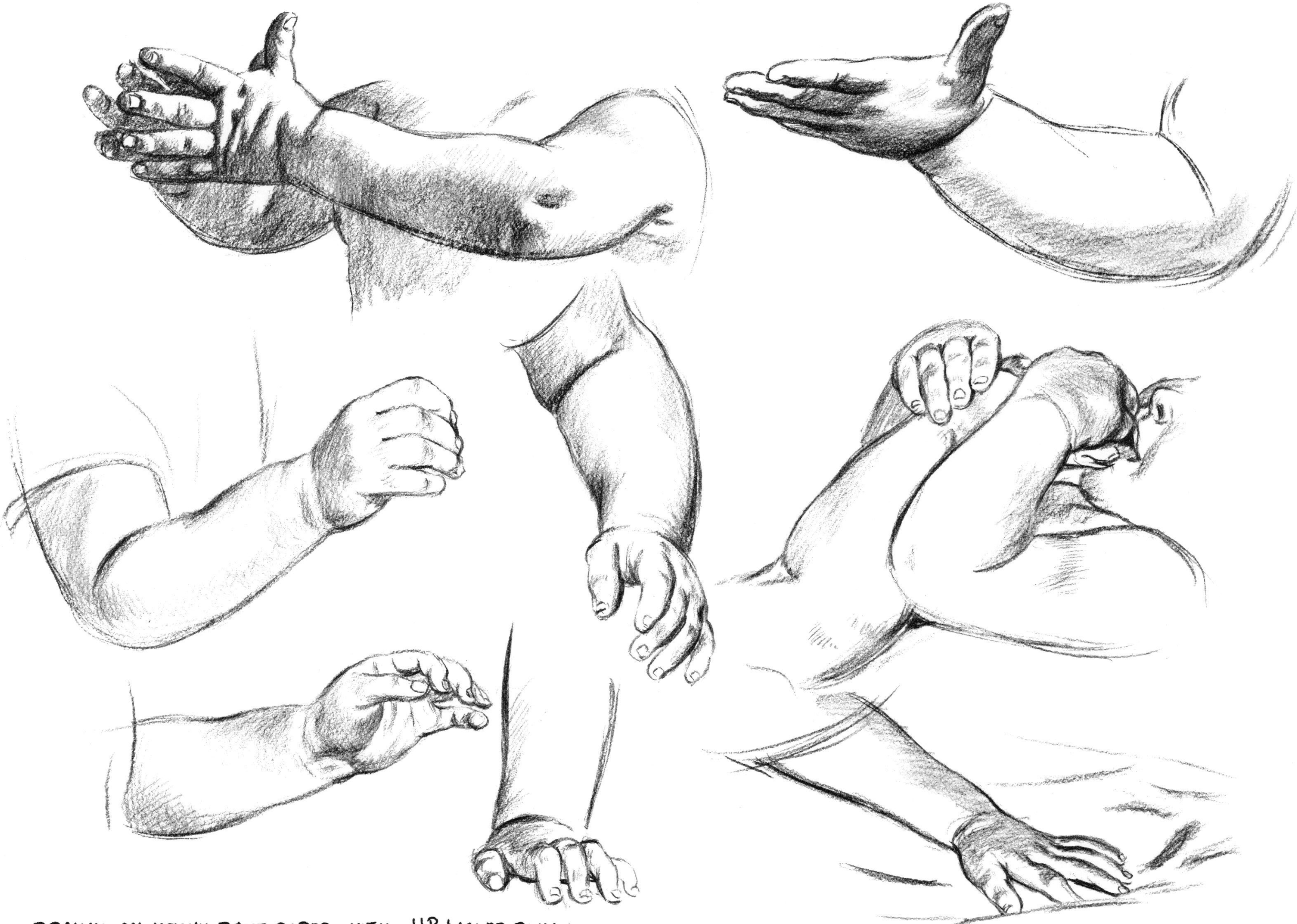

DRAWN ON HEAVY BOND PAPER WITH HB WOLFF PENCIL.

PRELIMINARY DRAWINGS & SKETCHES

THE FOLLOWING thirty pages of drawings have been included at the suggestion of the publisher. Over fourteen hundred drawings of hands along with several hundred studies of the anatomy were made by the author prior to the execution of the final plates. Obviously, there were many drawings that were made of the same action before one of them was chosen for inclusion in the book.

These following plates were the "thinking paper" of the author and were done simply as plans for finished drawings for the book. They were, in the author's mind, never intended to be published. The author has ambivalent feelings about their being a part of this book — however, he would never question the sagacity of his publisher — heaven forbid.

Some of the plates have written upon them reminders to the author that were hastily jotted down while he was making the drawings. On one of the plates the reader shall note, "1.26 shake 3 seconds 1:30 2nd shake 5 seconds" and then lower on the page, "1:39 real jolt". These cryptic notes could be read several ways if one did not know what they meant; for instance, one might take them as directions for the mixing of some exotic cocktail that demanded highly controlled shaking. They are, however, the recording of the time of the after-tremblors of the tragic Sylmar earthquake. As can be seen by the drawing above the notation, "1:39 real jolt", the author ceased work and *did* mix himself a drink. This business of earthquakes is one of the hazards one faces as a resident of Southern California.

The author reluctantly admits that some of these preliminary drawings are better than his finished drawings — but then again whose sketches are not often better than his finished works? It is something like our speech — in front of those we don't know we often make rather stilted statements while among friends we are less on guard and, therefore, speak in a freer manner — sketches have the latter quality. Gentle reader, look upon these plates as having been drawn before friends — for you are friends; after all you did buy the book. For this latter act, the author thanks you, his family thanks you and most assuredly his publisher thanks you.

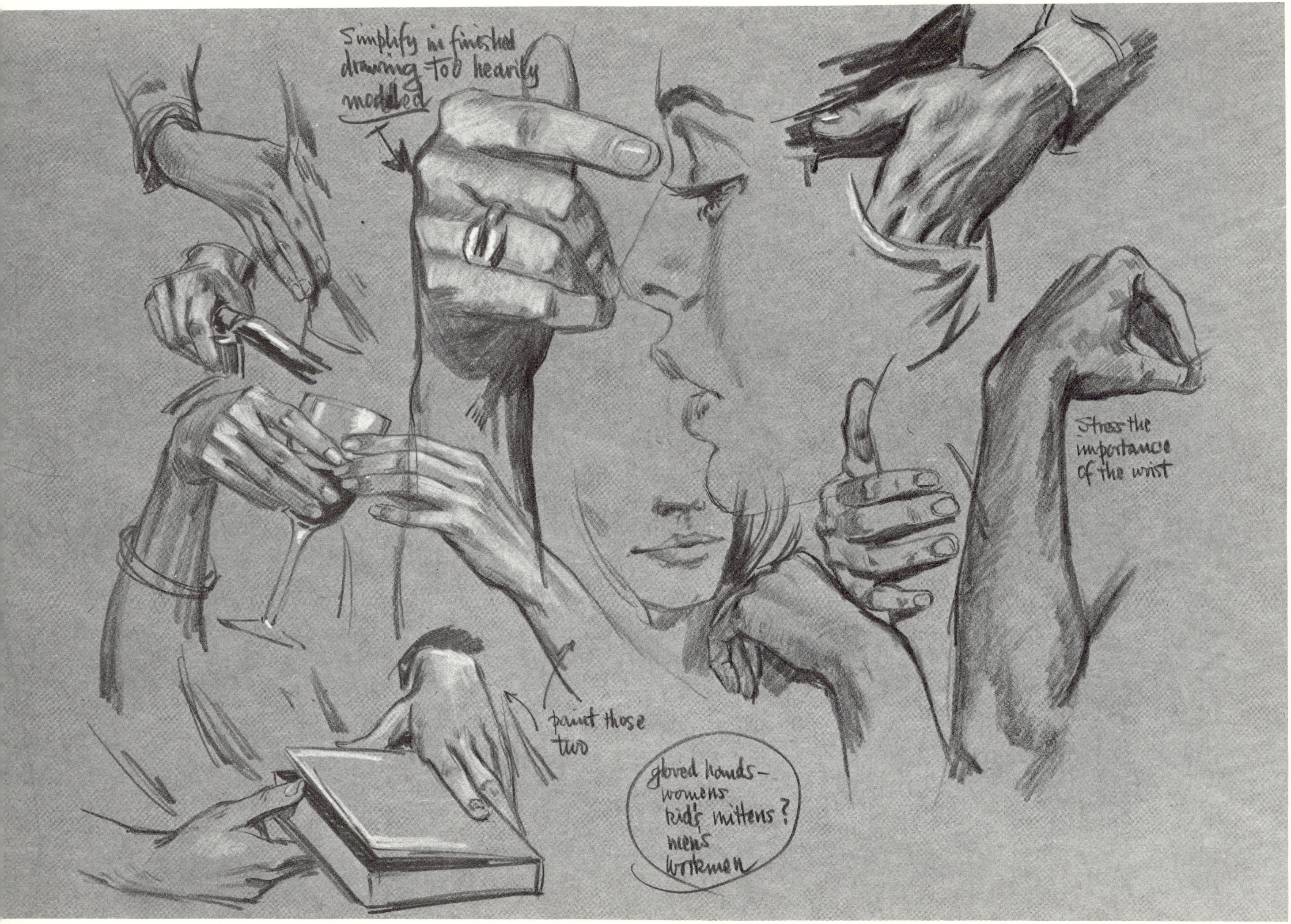
Simplify in finished drawing-too heavily modelled
Stress the importance of the wrist
paint those two
gloved hands-
womens
kid's mittens?
men's
workmen

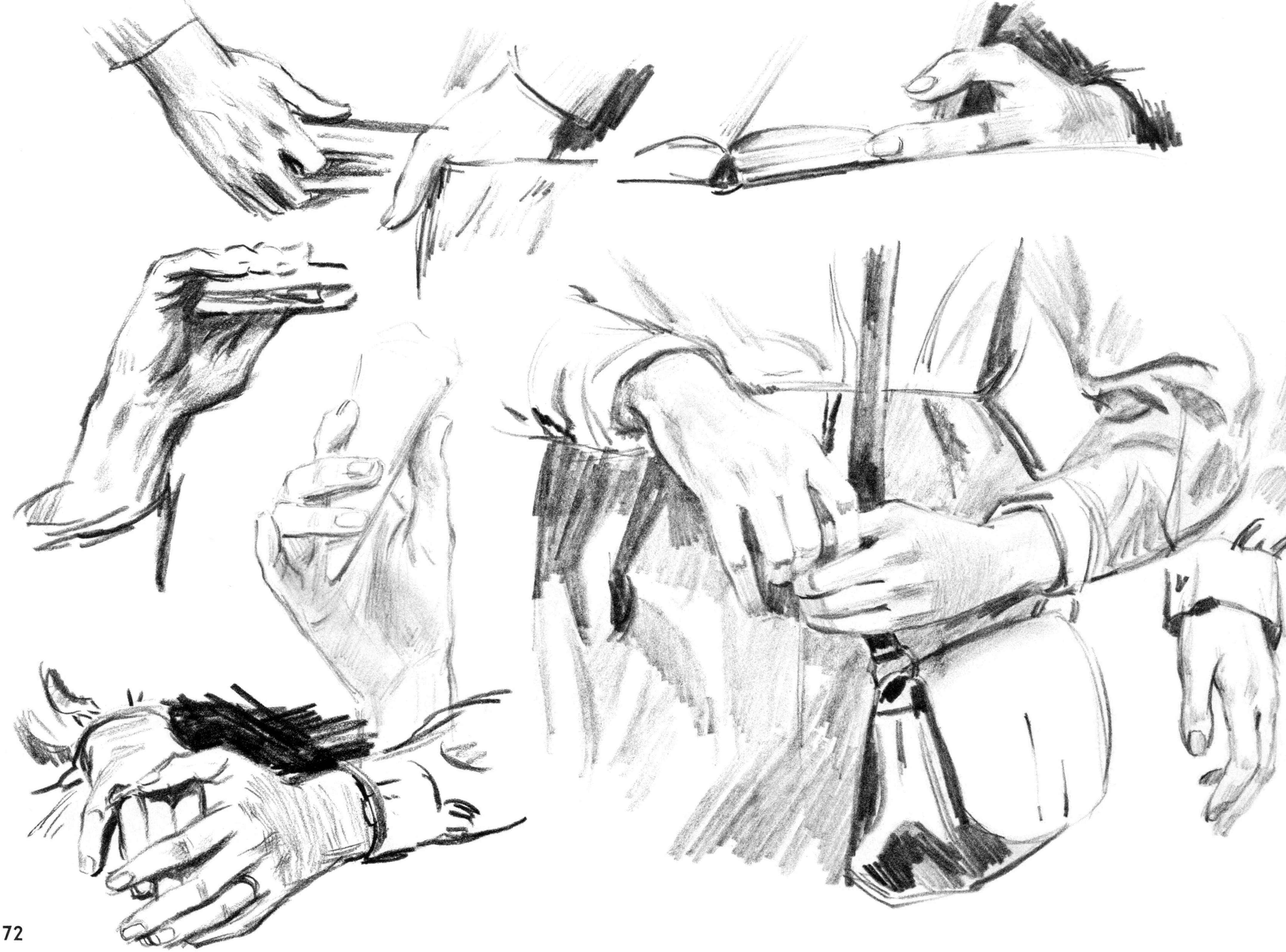

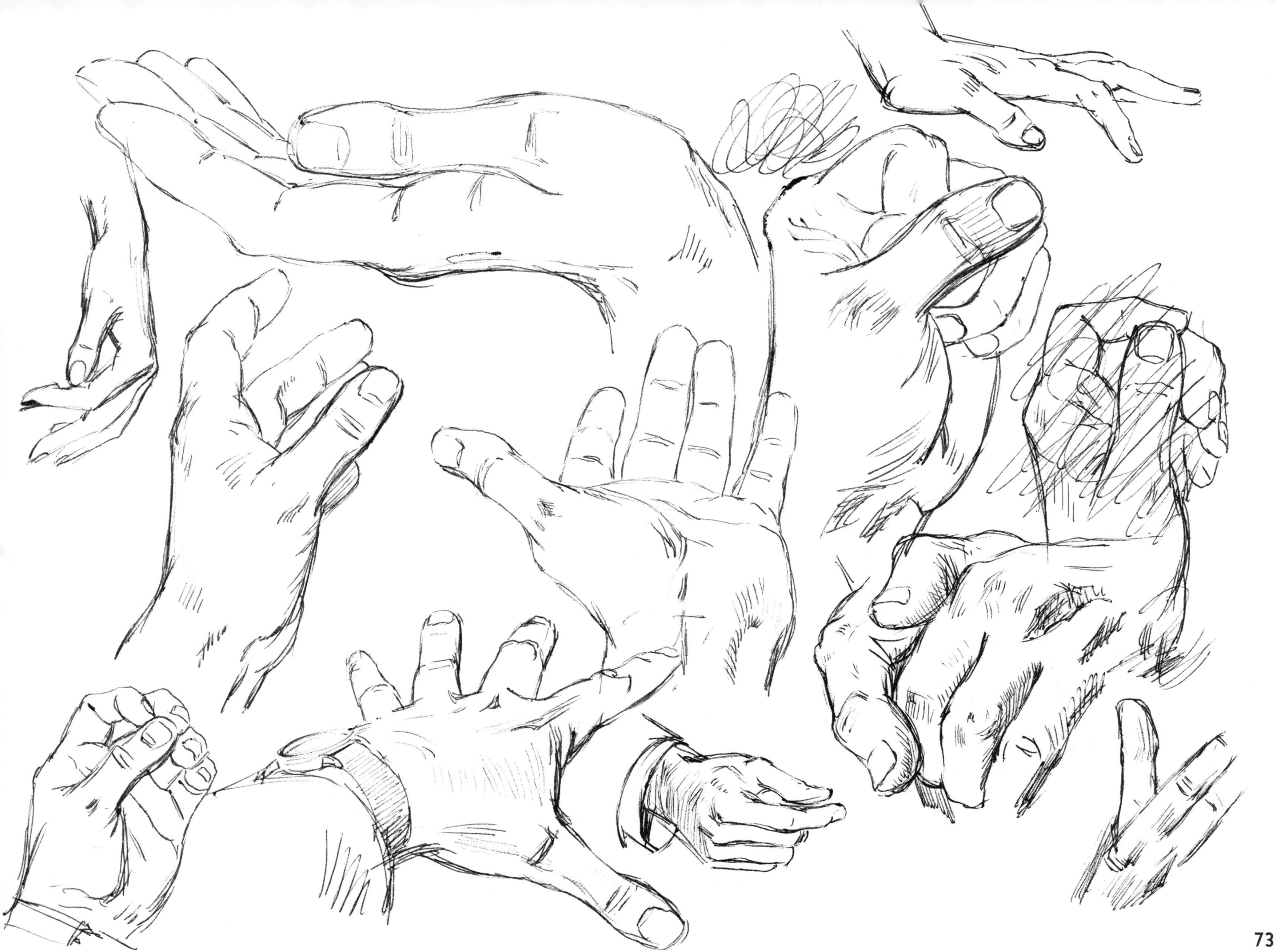

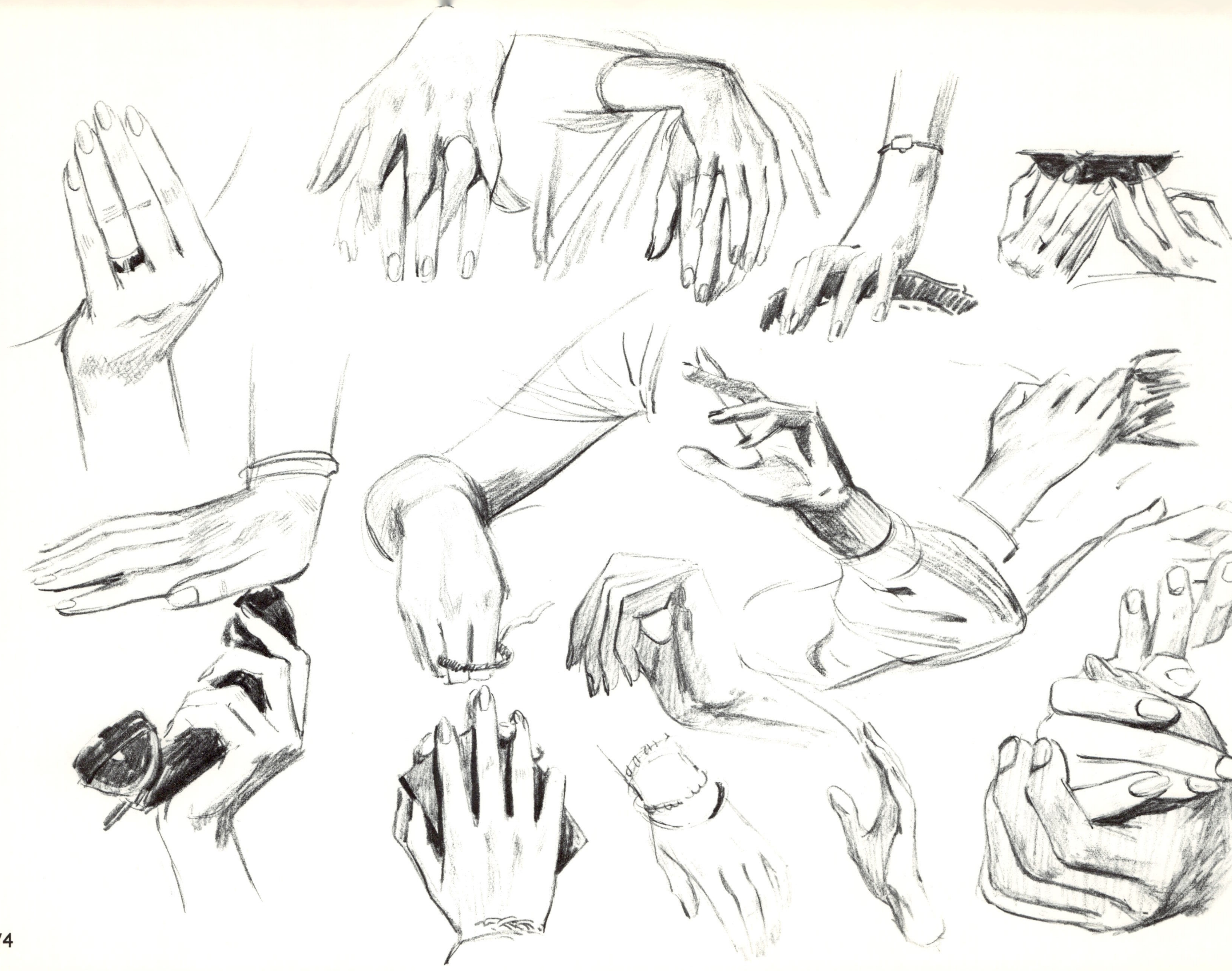

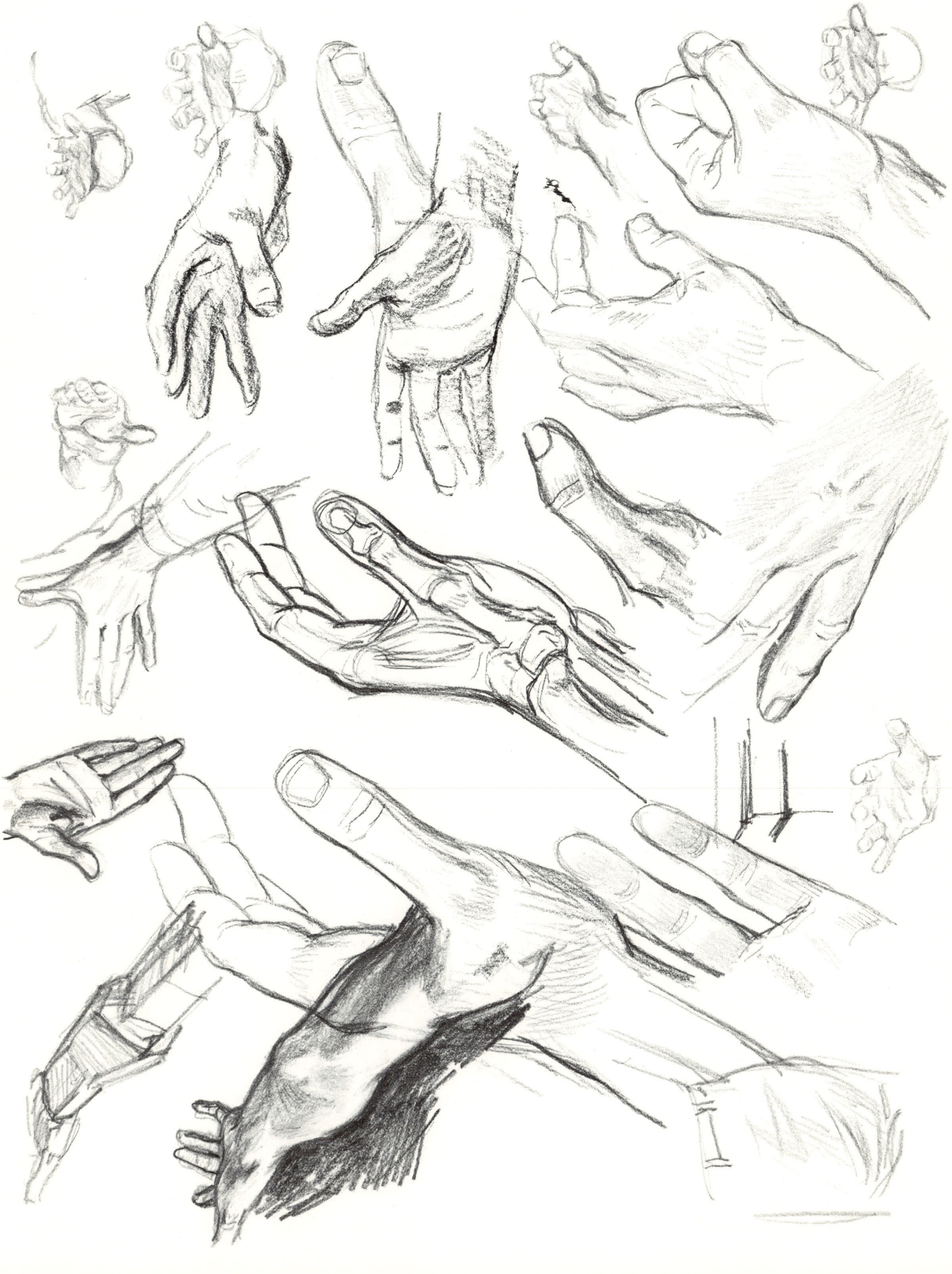

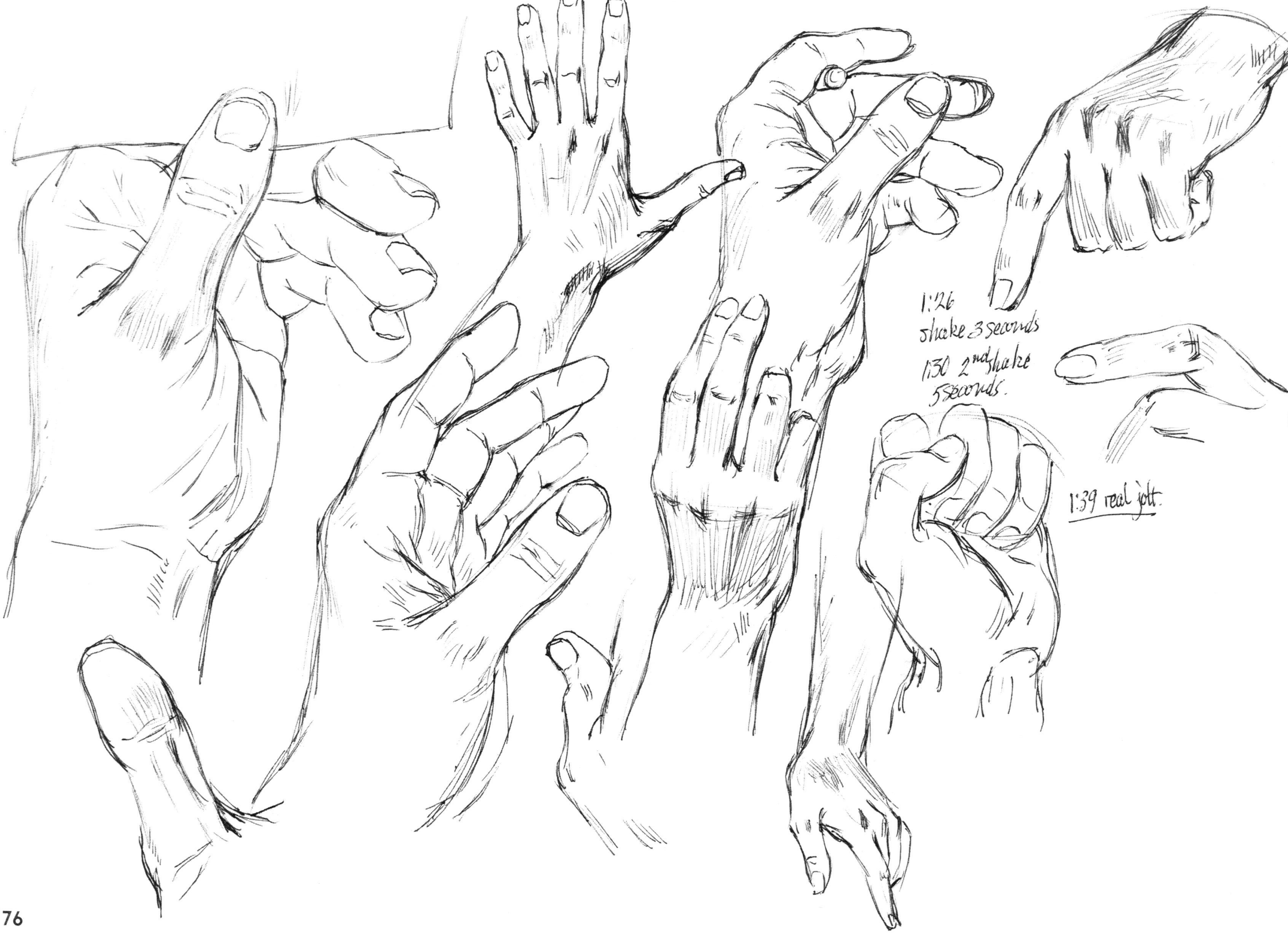
1:26
shake 3 seconds
1:30 2nd shake
5 seconds.
1:39 real jolt.

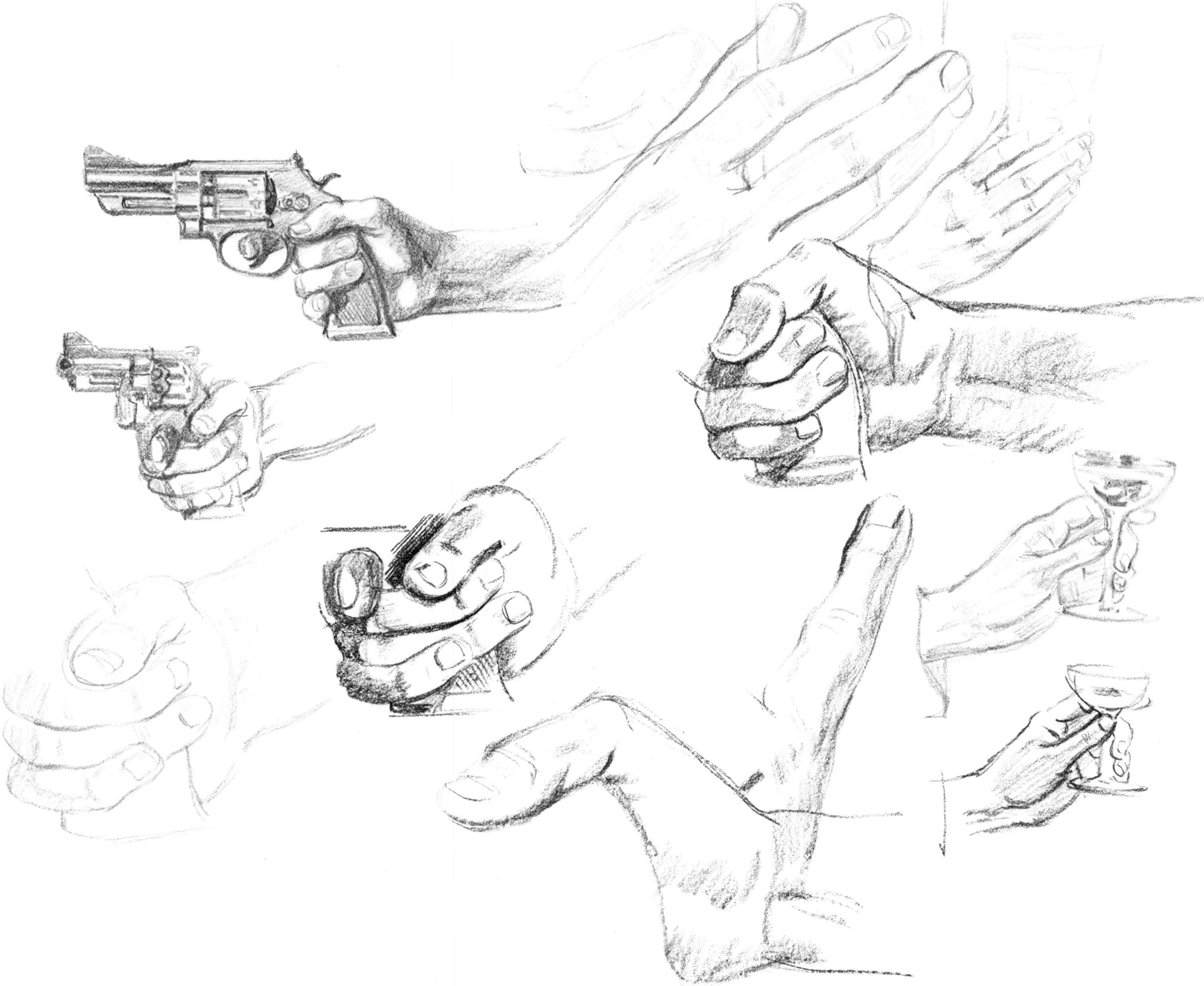

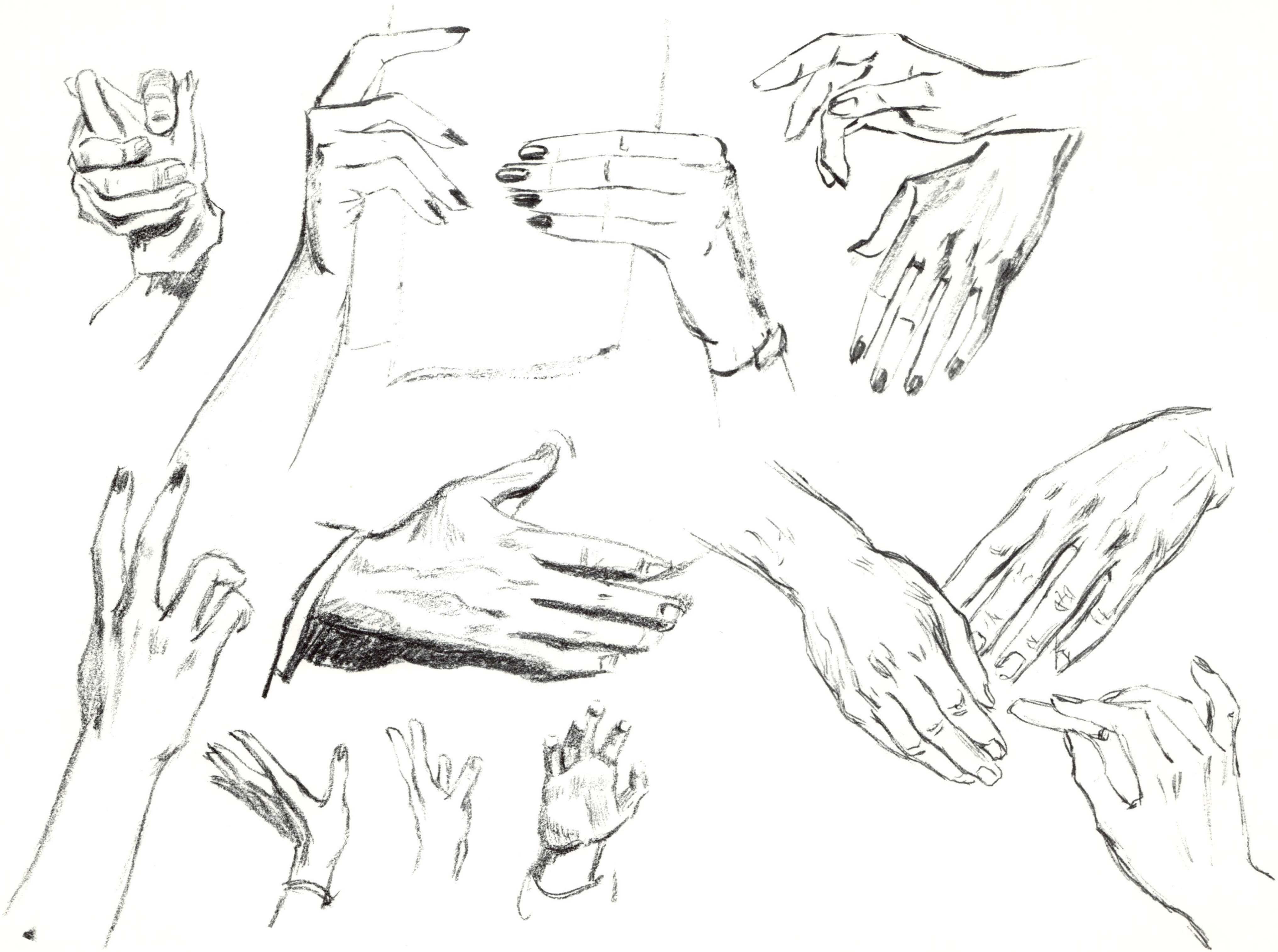

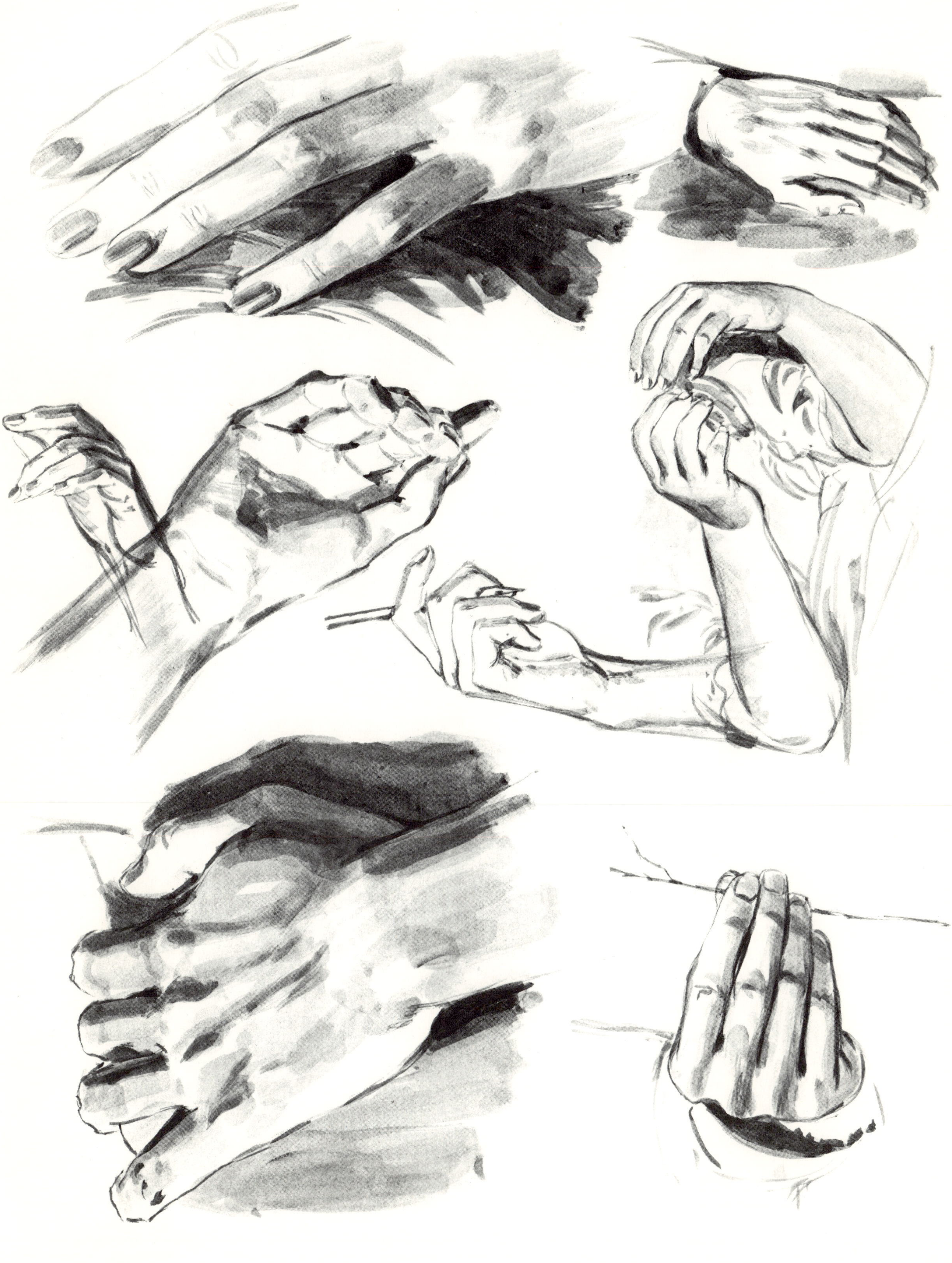

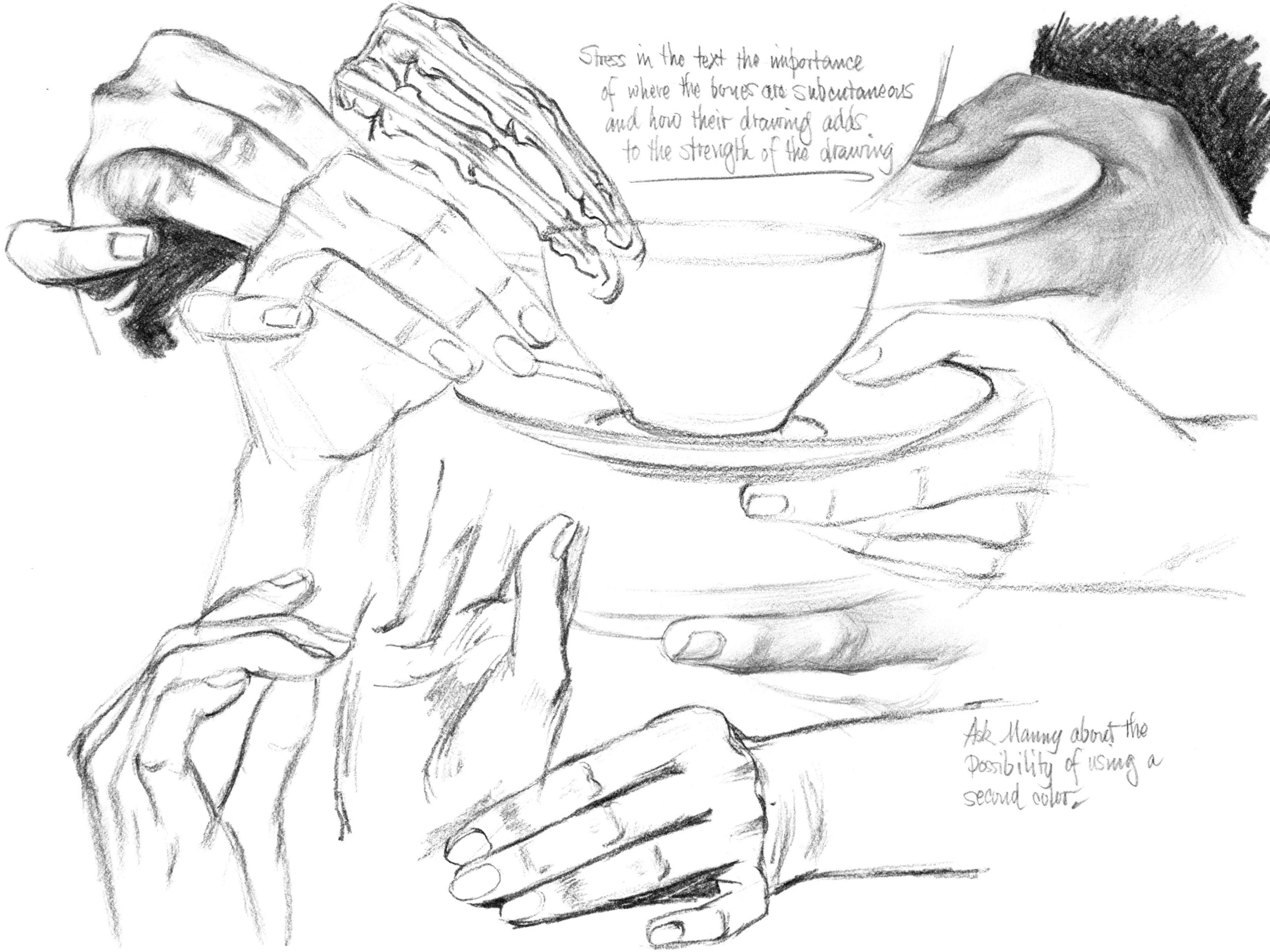
Stress in the text the importance
of where the bones are subcutaneous
and how their drawing adds
to the strength of the drawing
Ask Manny about the
Possibility of using a
second color.

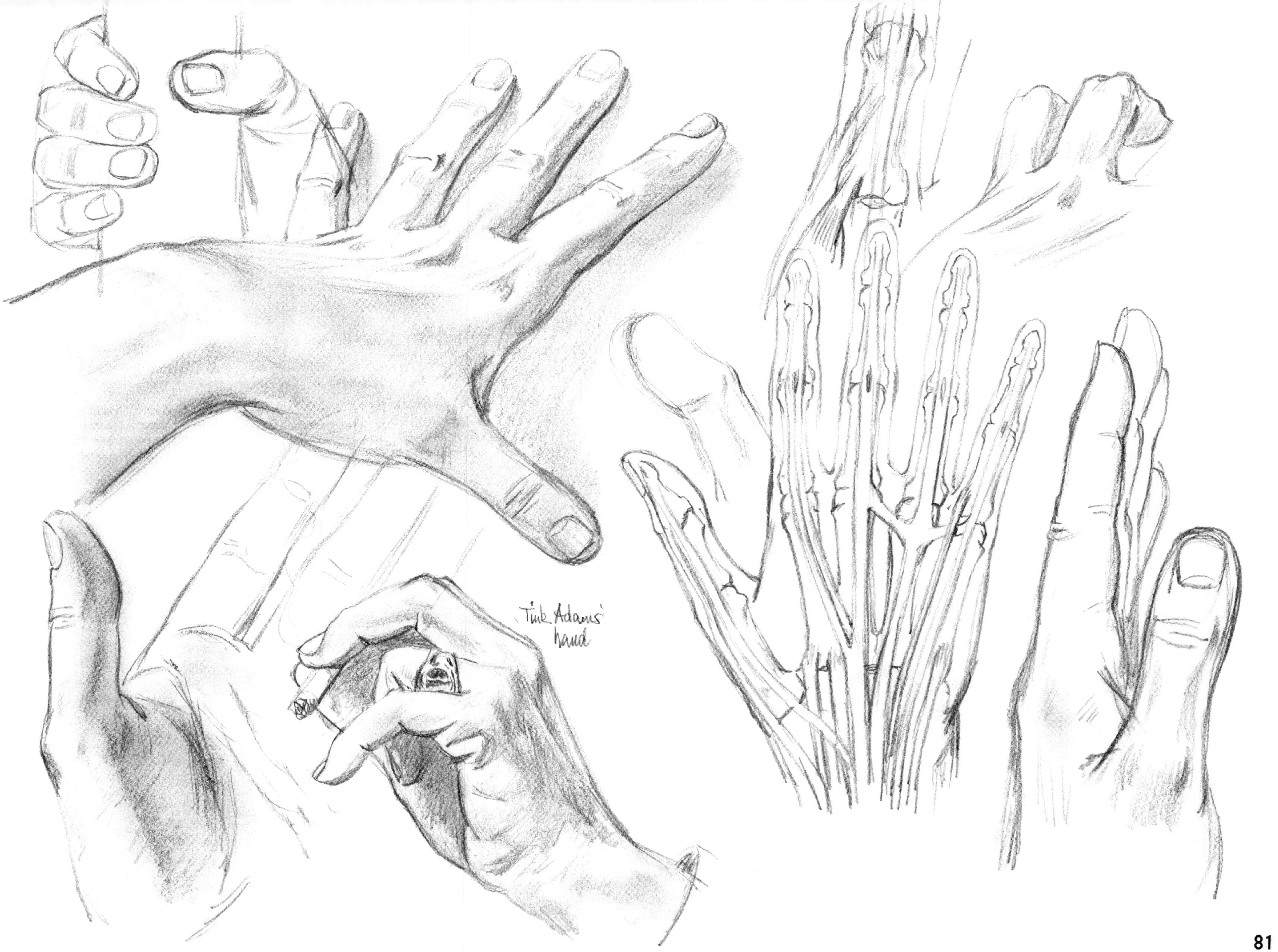
Tink Adams' hand

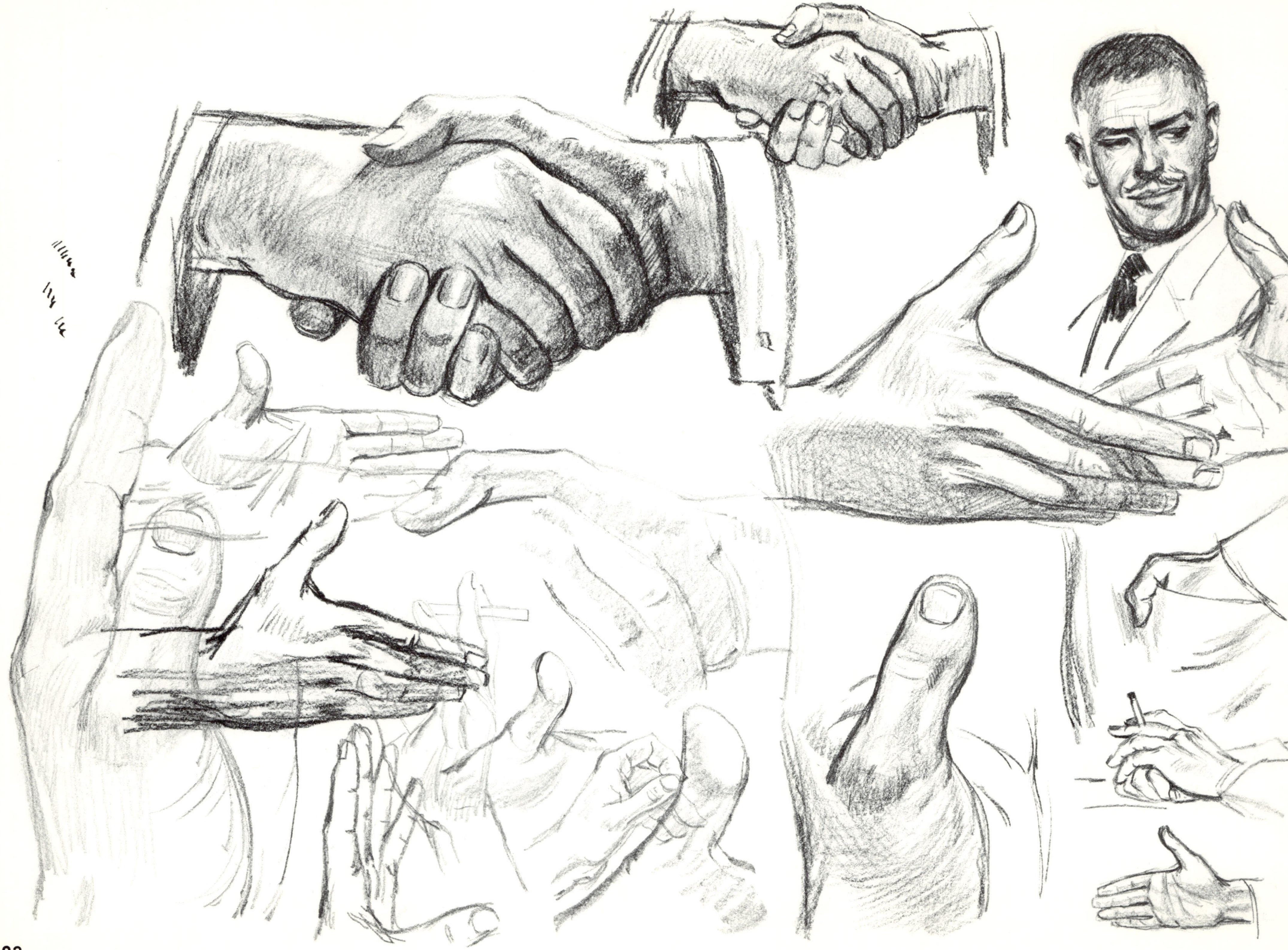

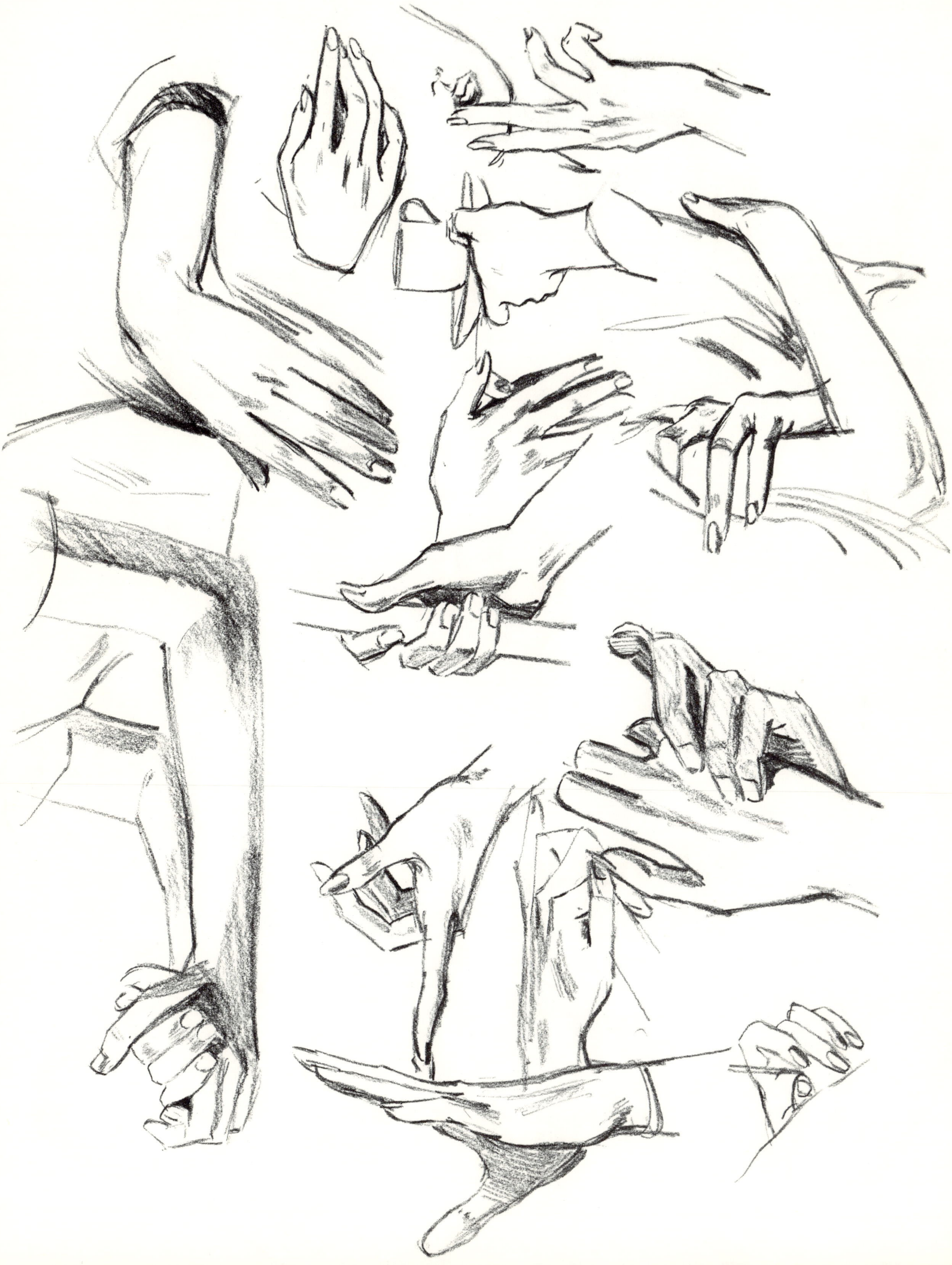

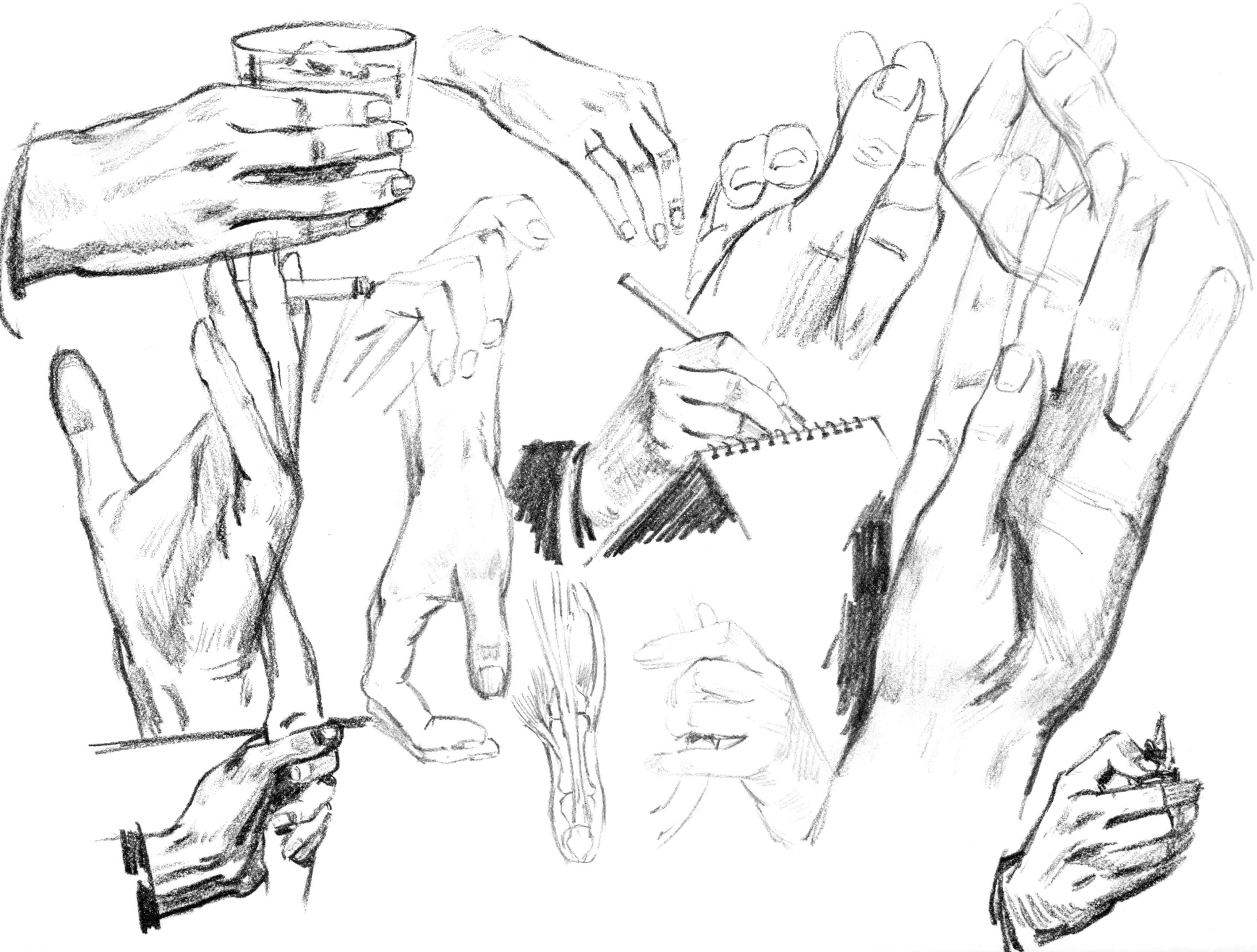

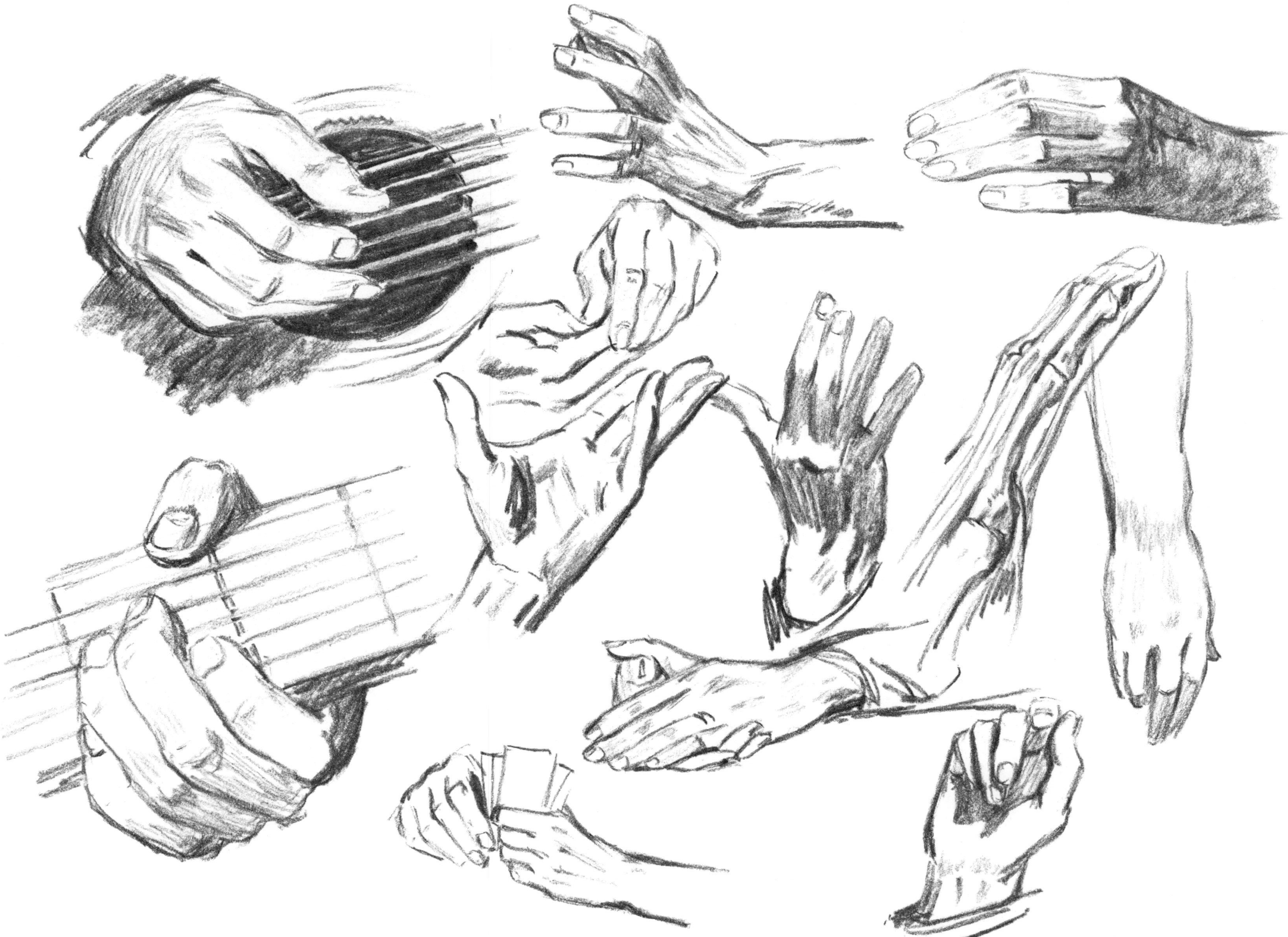

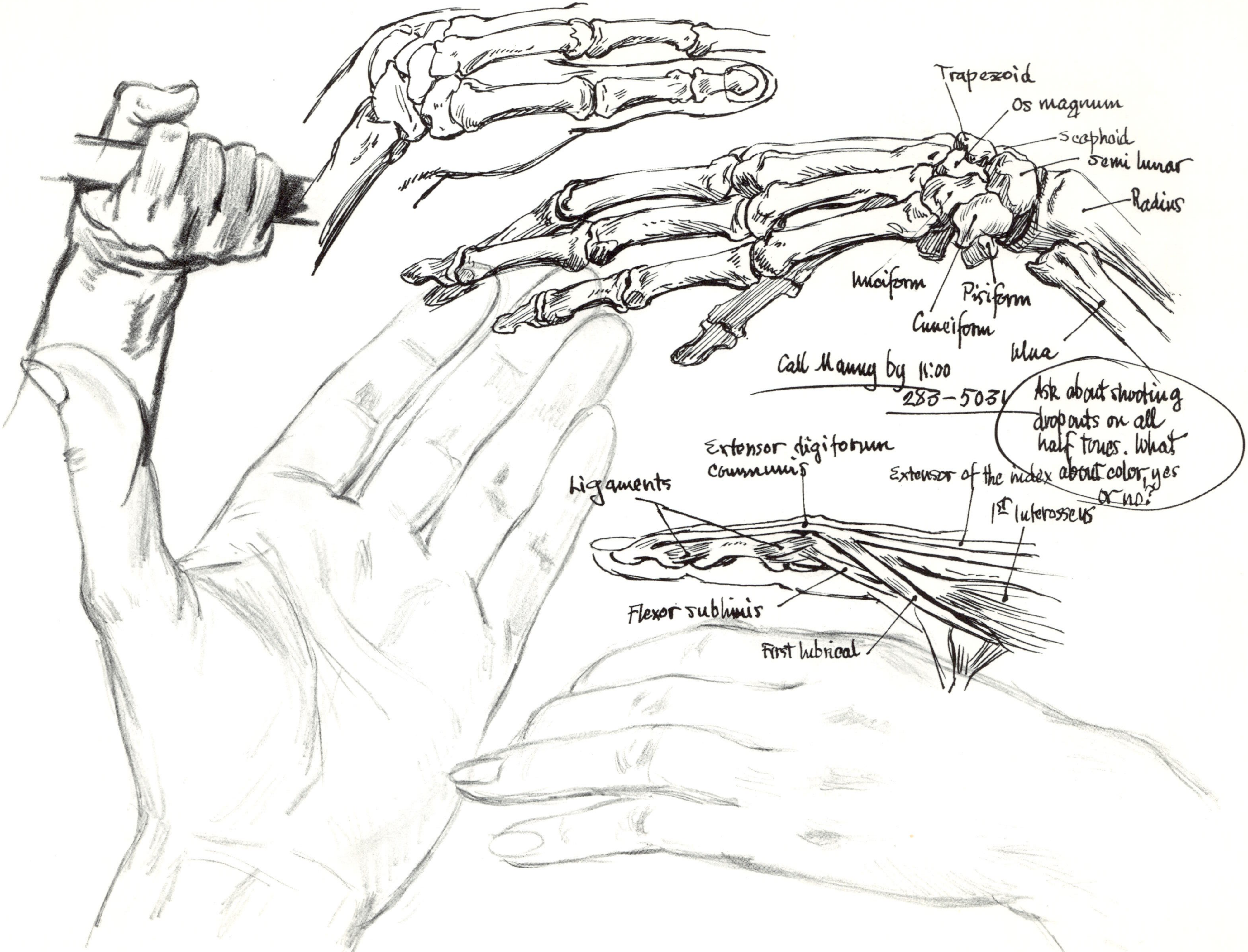
Trapezoid
Os magnum
Scaphoid
Semi lunar
Radius
Unciform
Pisiform
Cuneiform
Ulna
Call Manny by 11:00
283-5031
Ask about shooting dropouts on all half tones. What about color, yes or no?
Extensor digitorum communis
Ligaments
Extensor of the index
1st Interosseus
Flexor sublimis
First lubrical

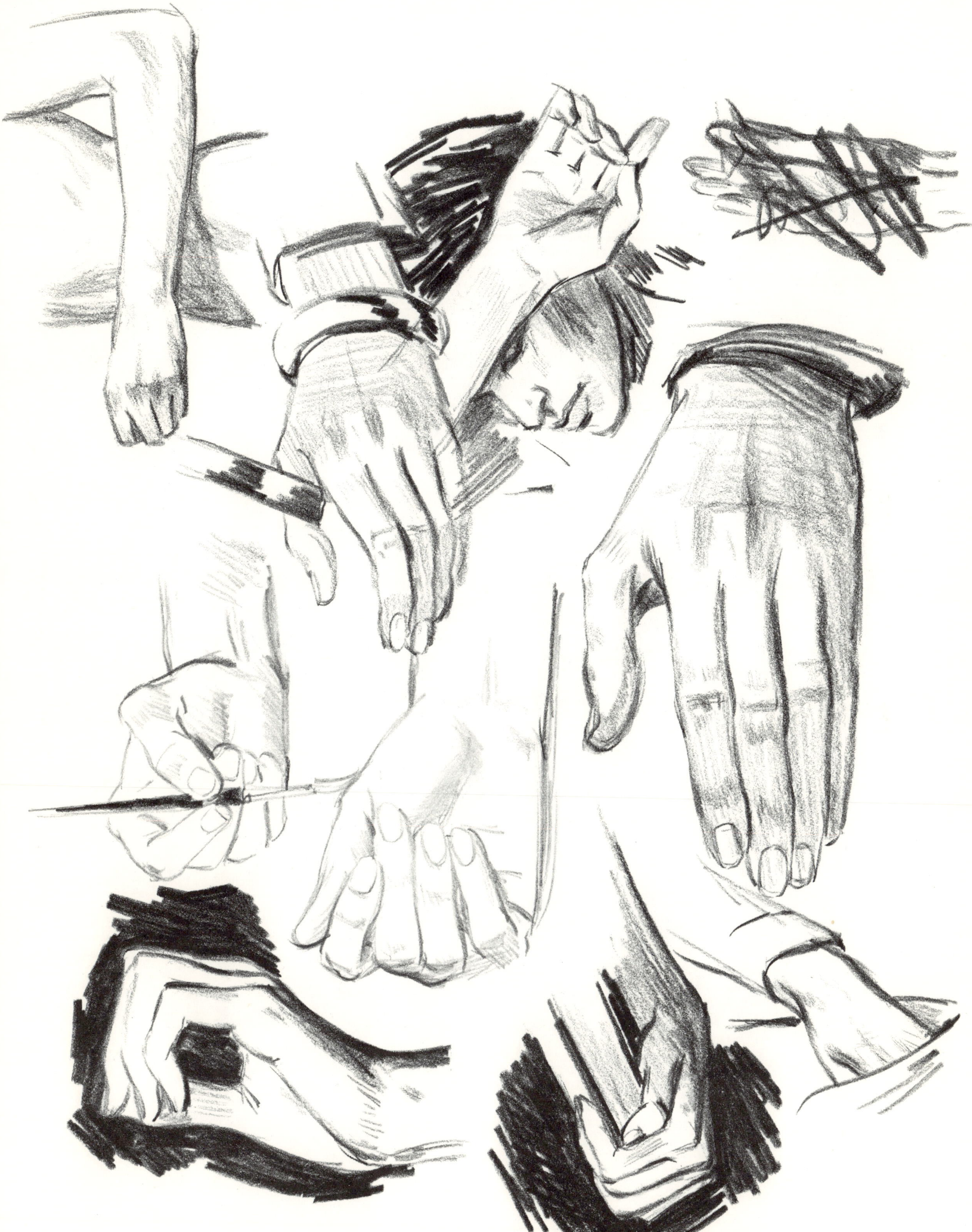

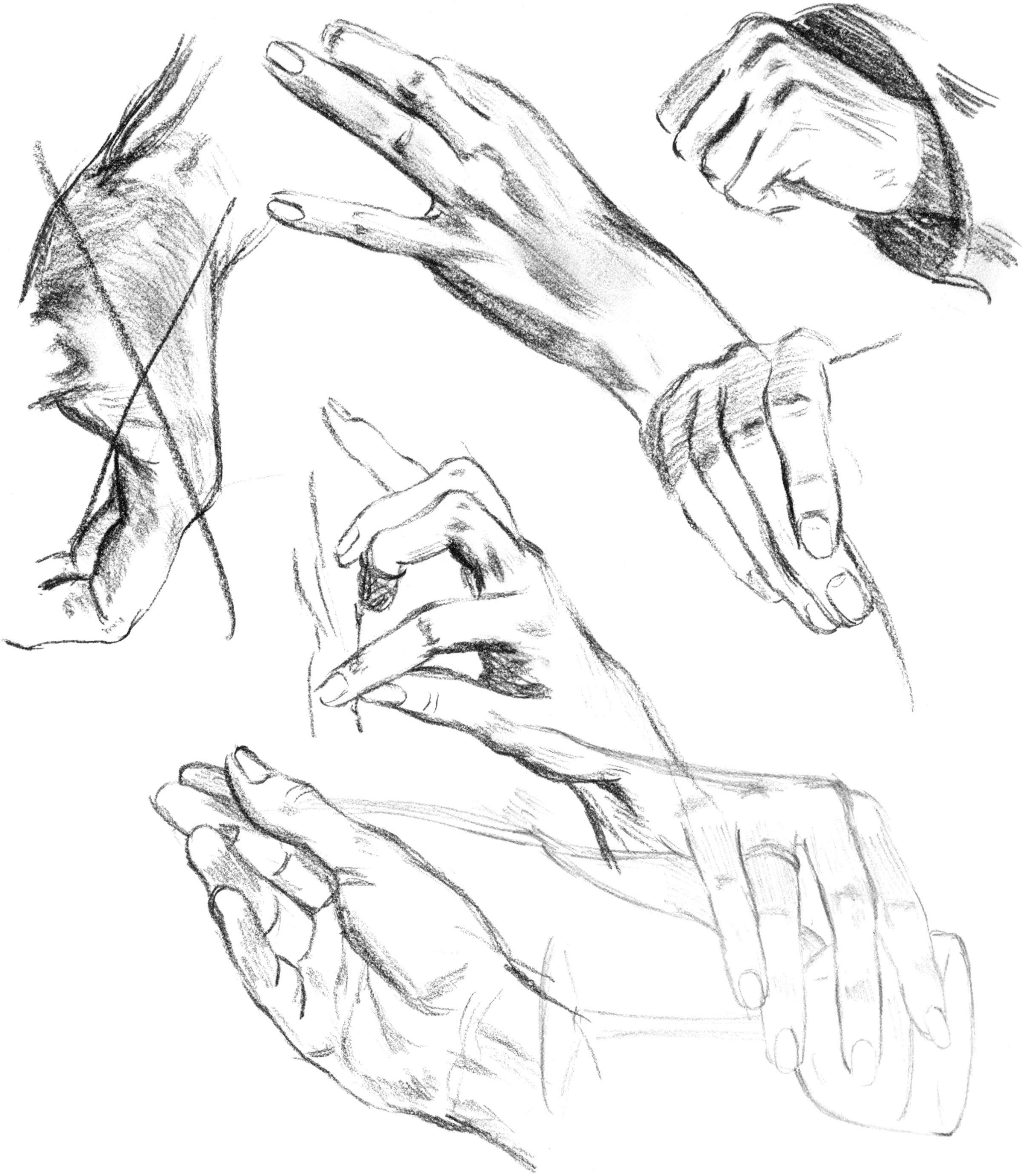

STUDIES OF HANDS FOR PORTRAITS.

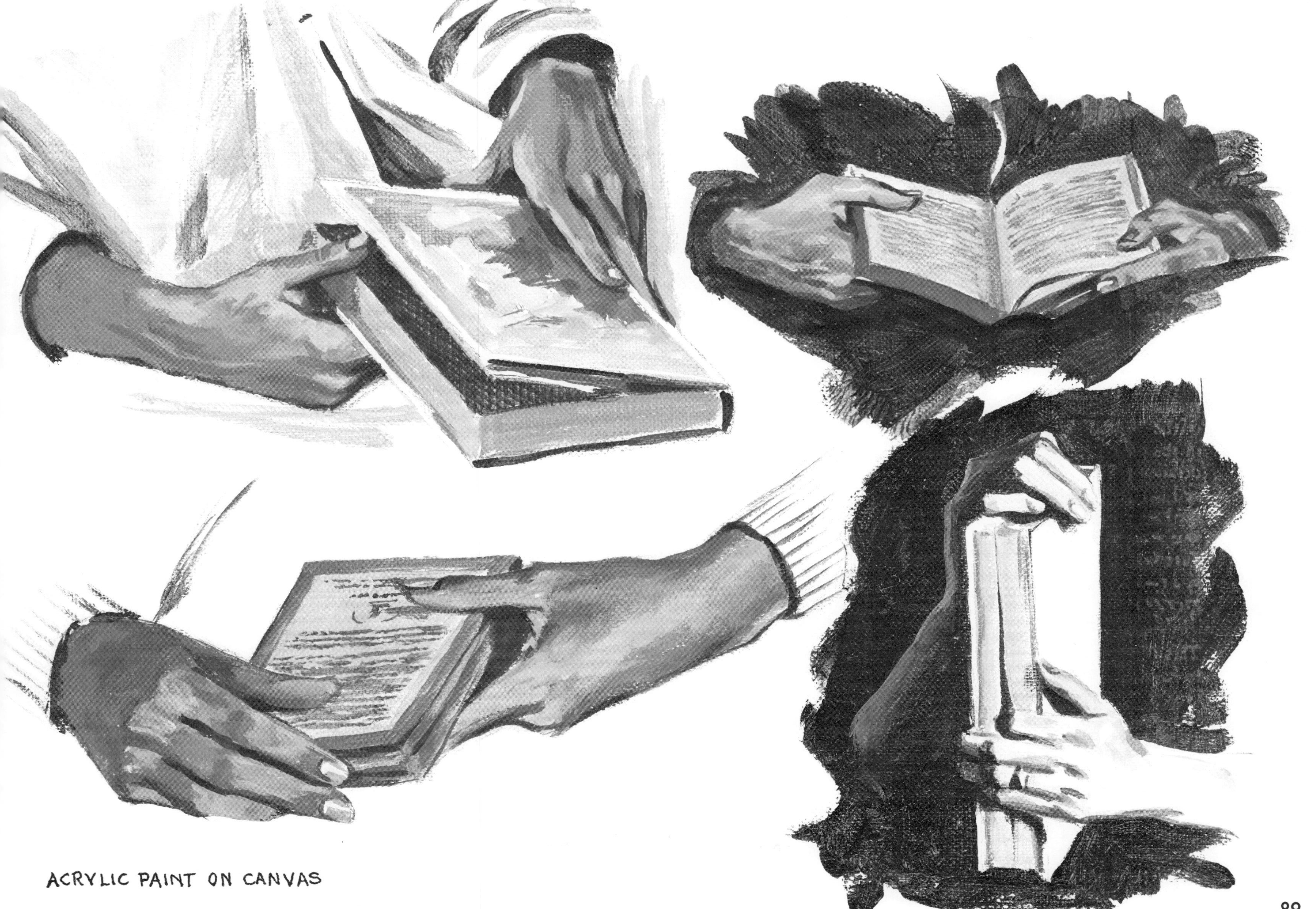

ACRYLIC PAINT ON CANVAS

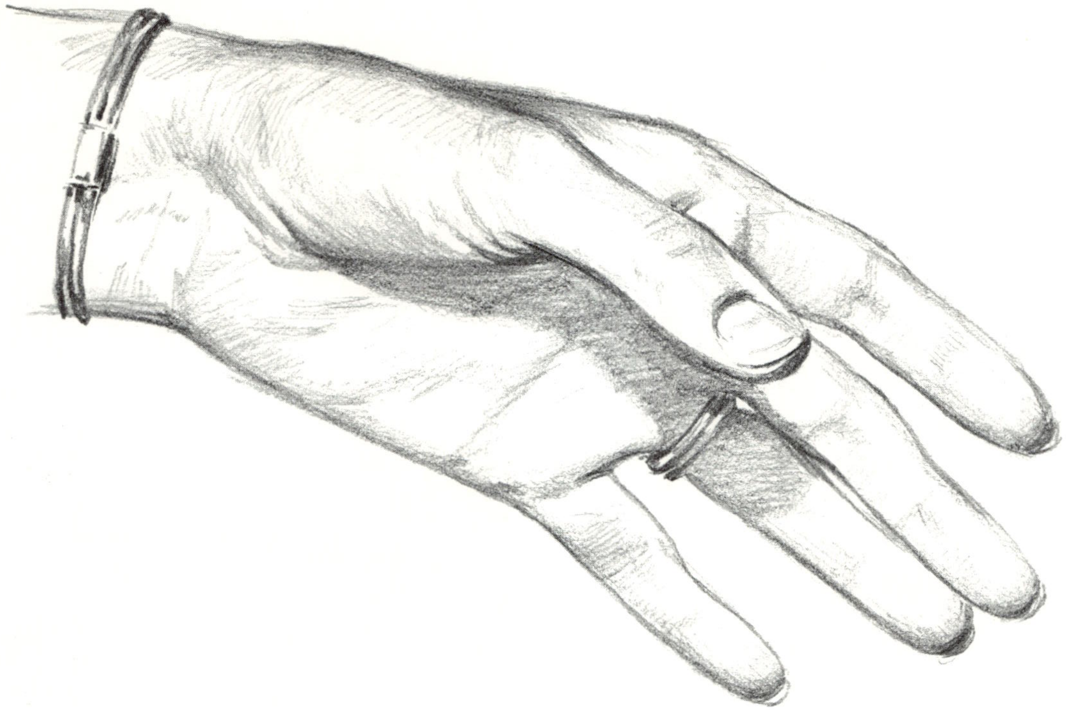

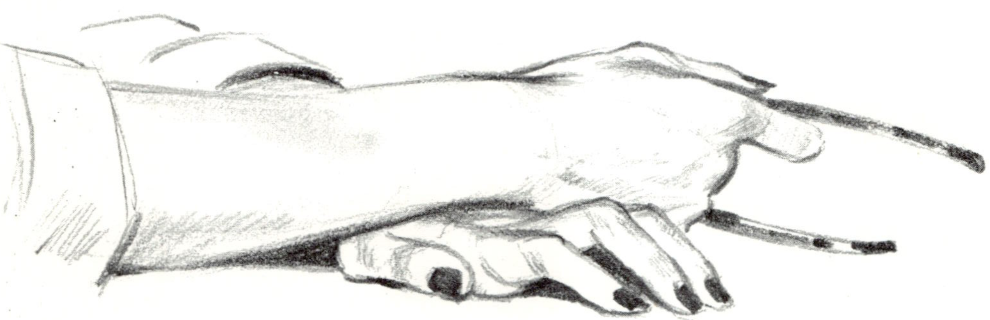

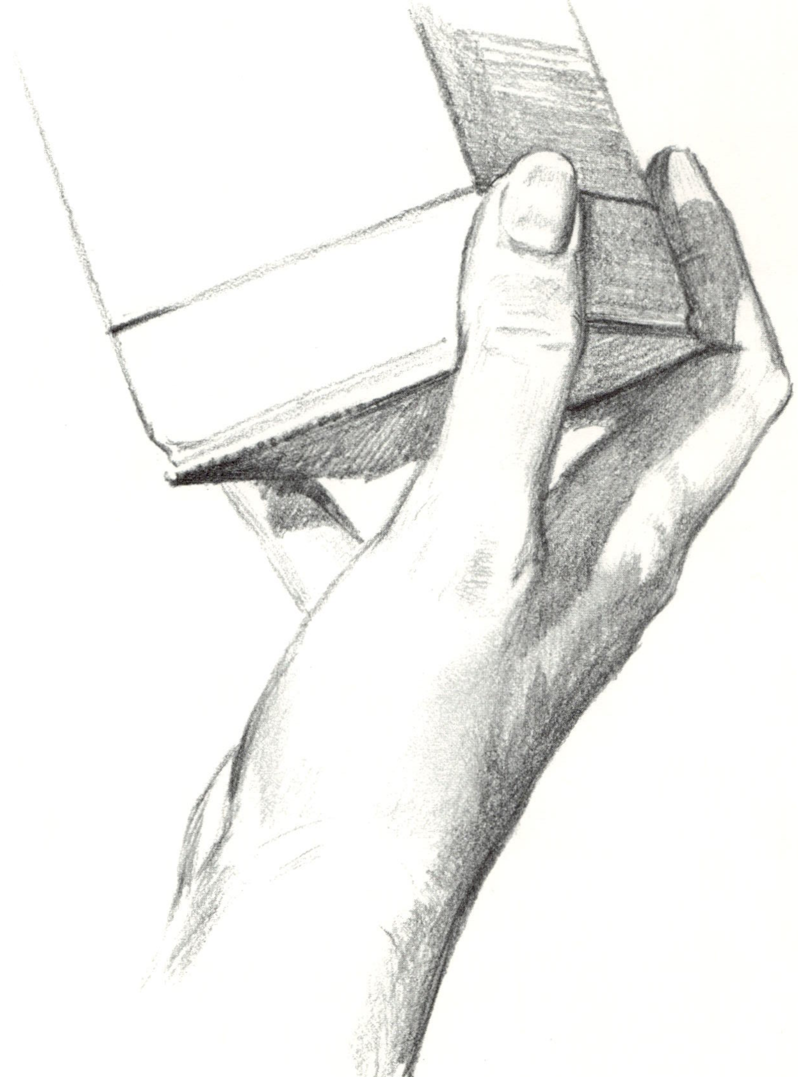

DRAWN ON HEAVY WEIGHT BOND PAPER WITH 2B GRAPHITE PENCIL.

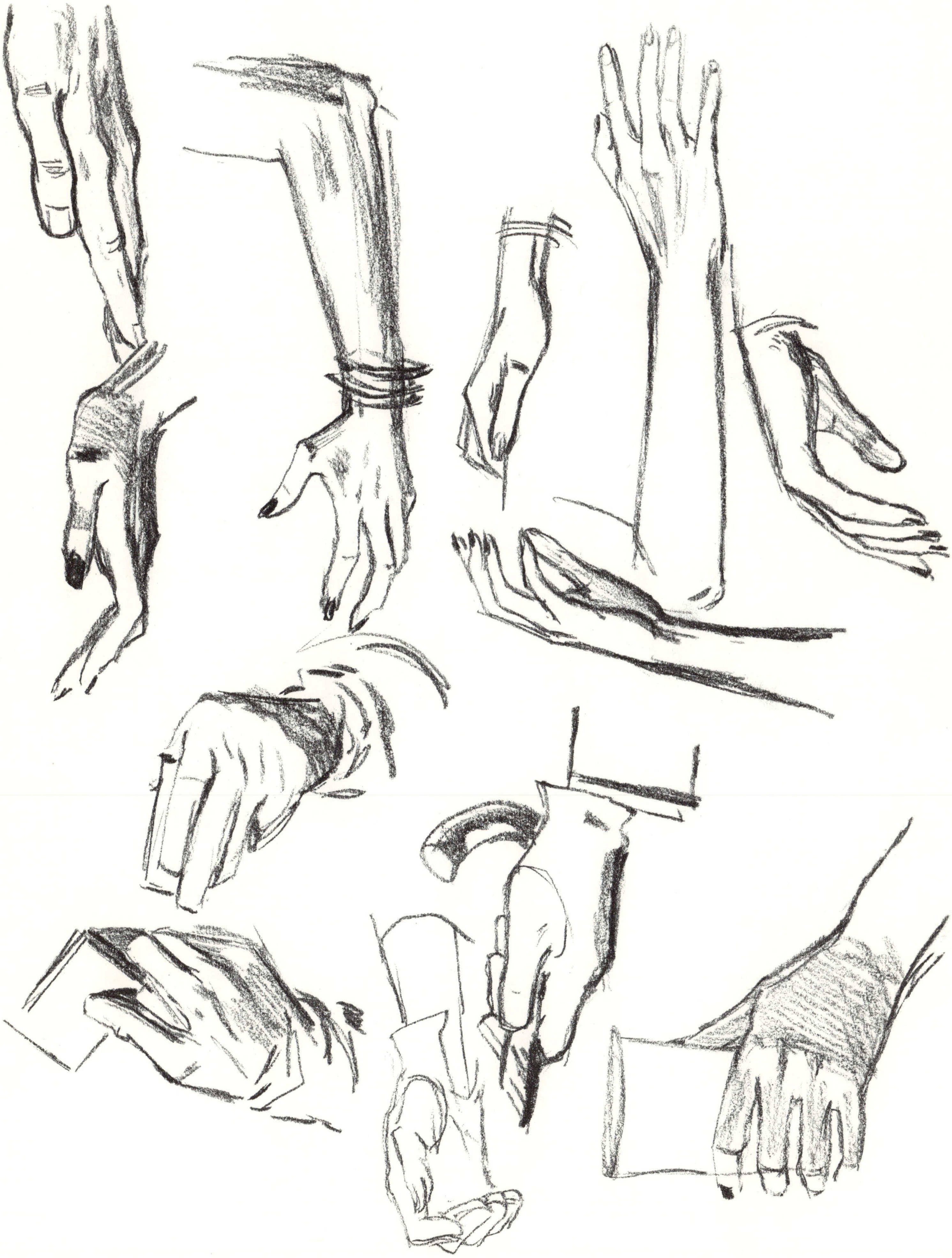

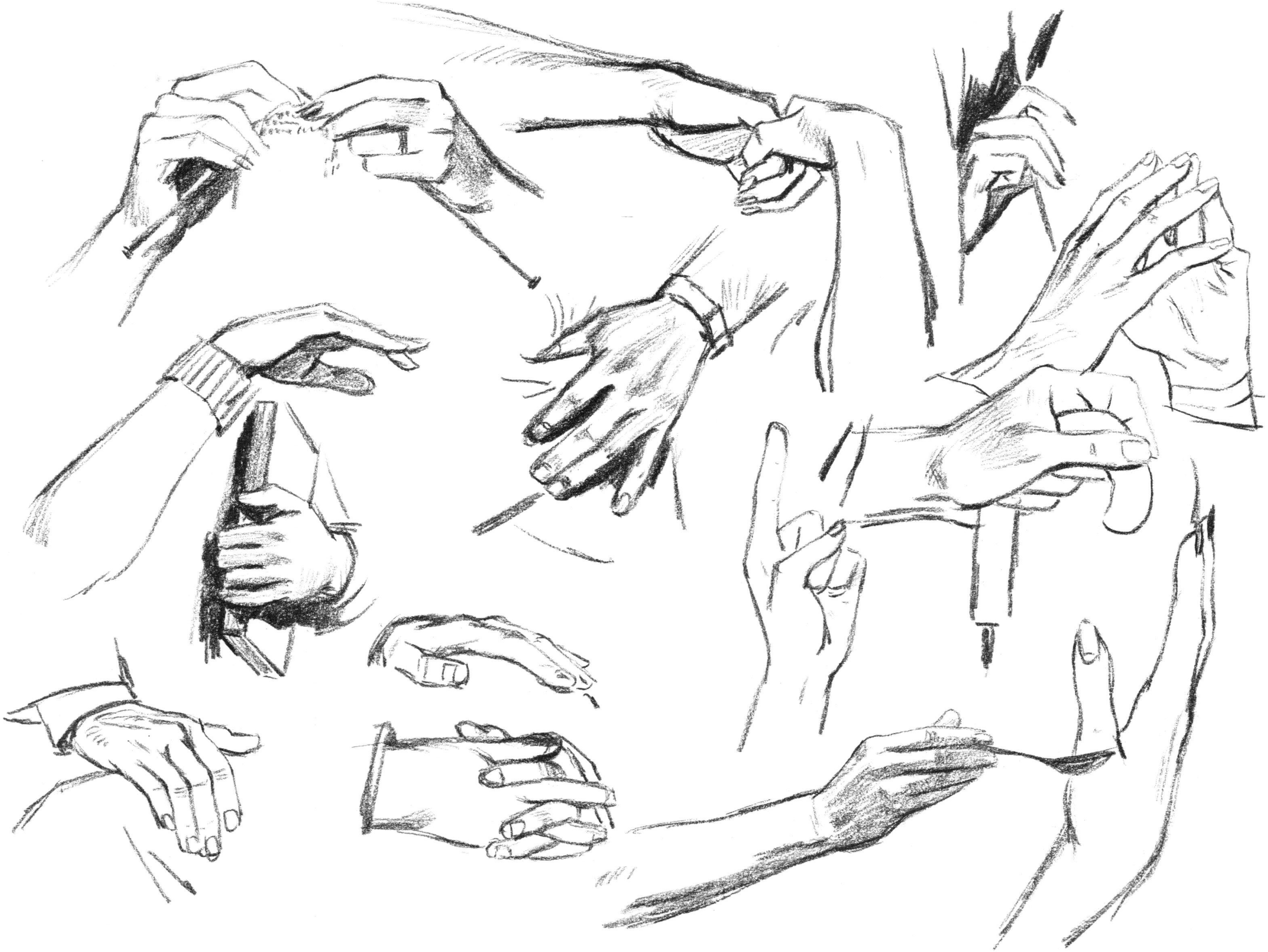

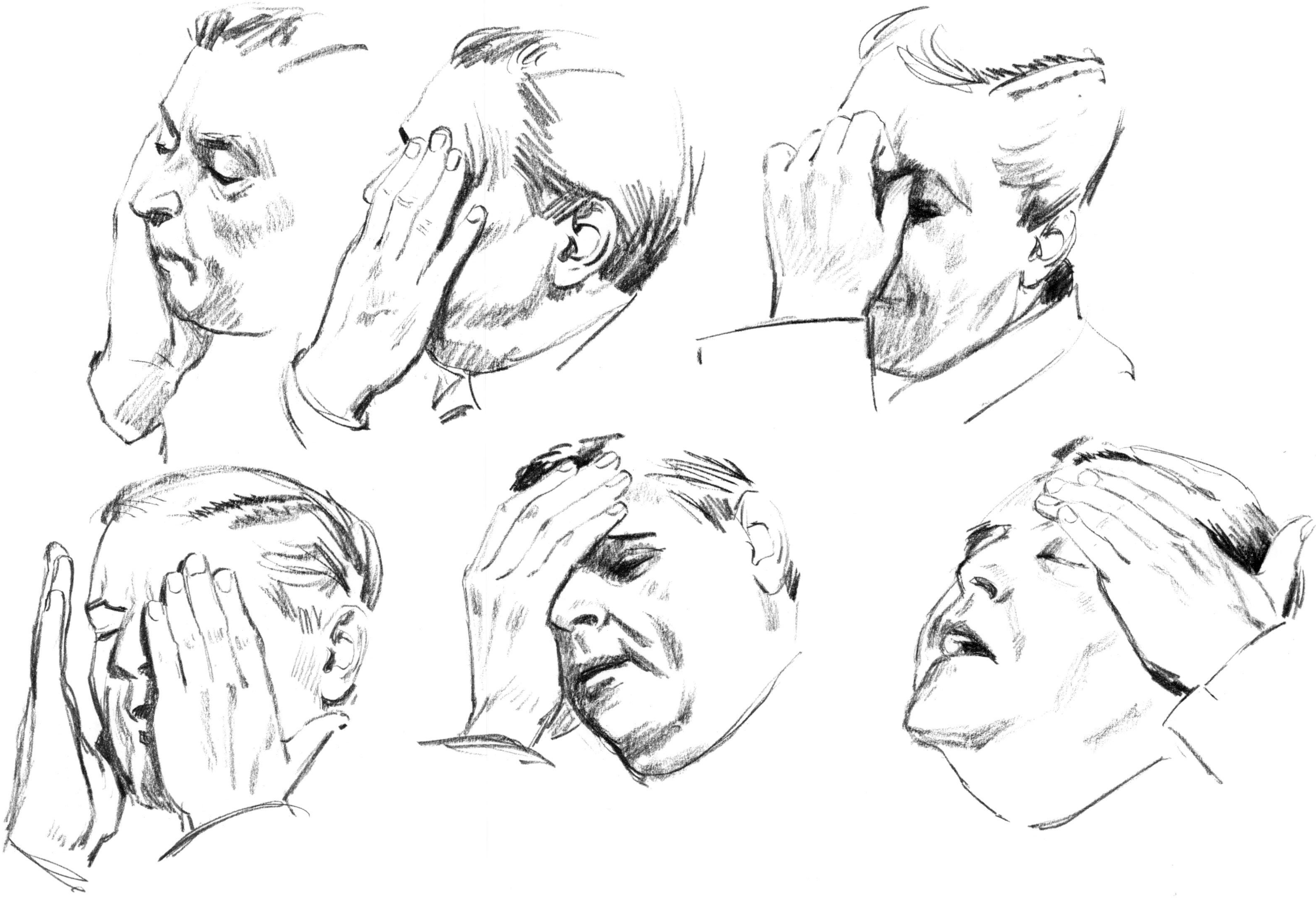

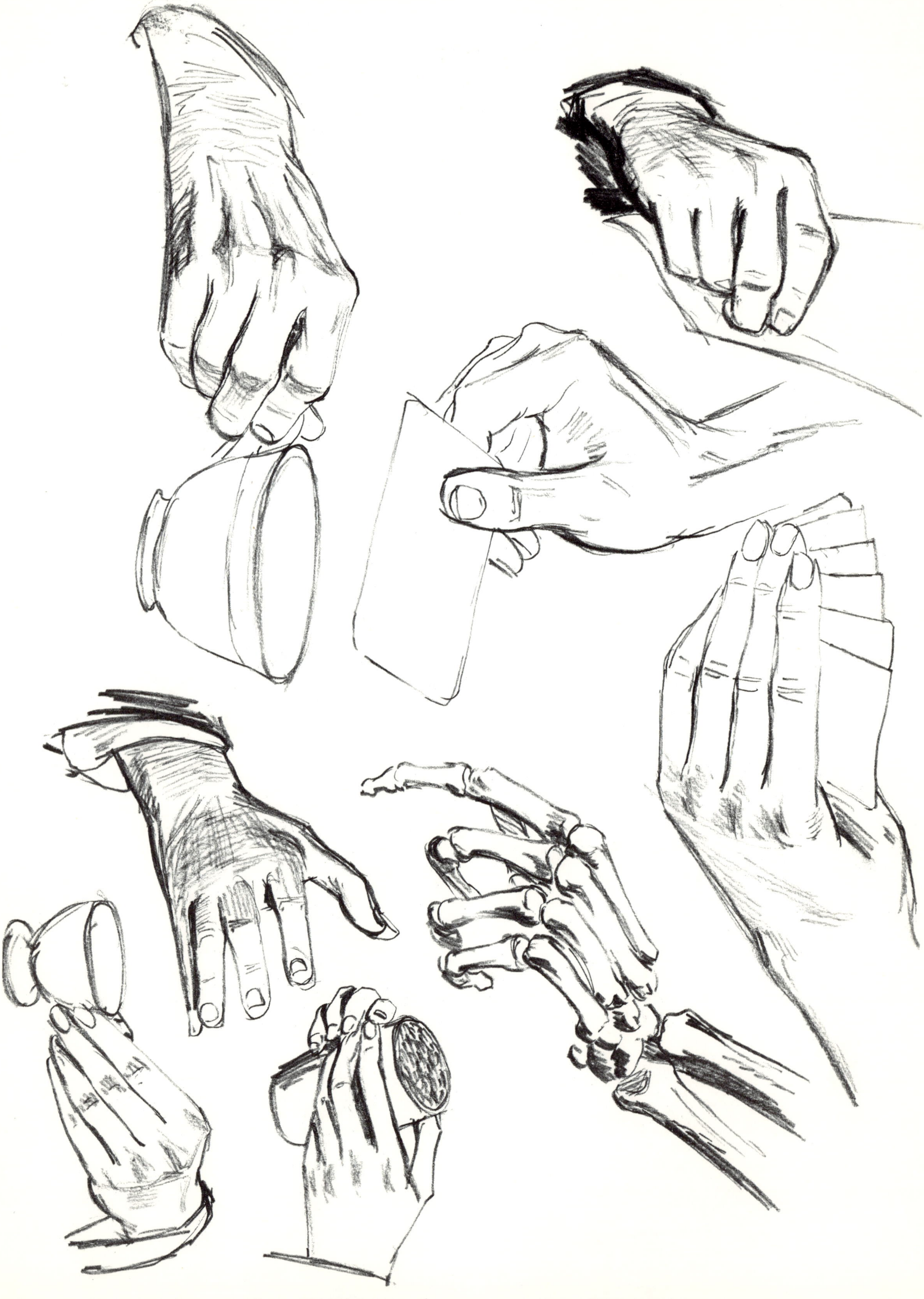

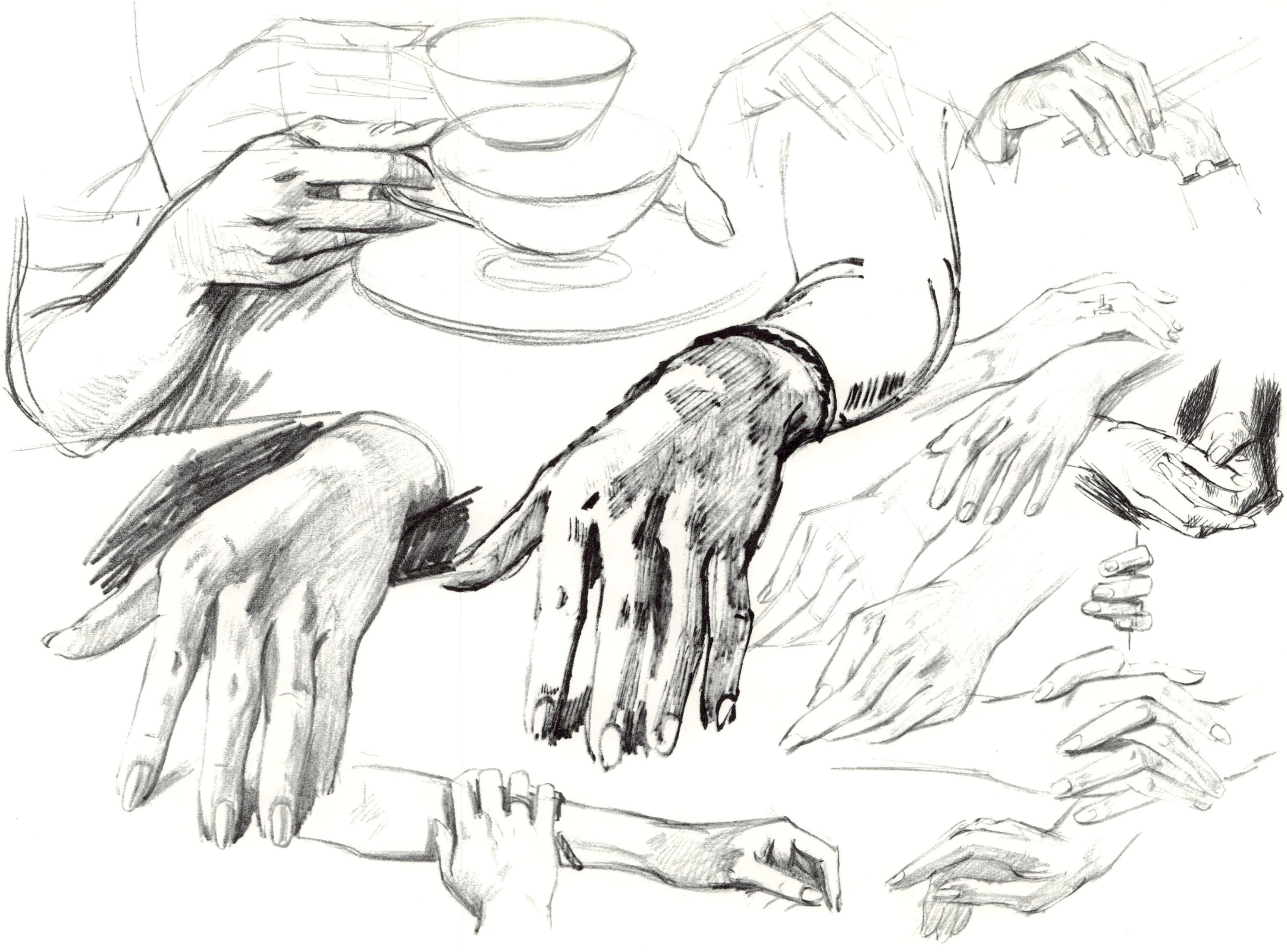

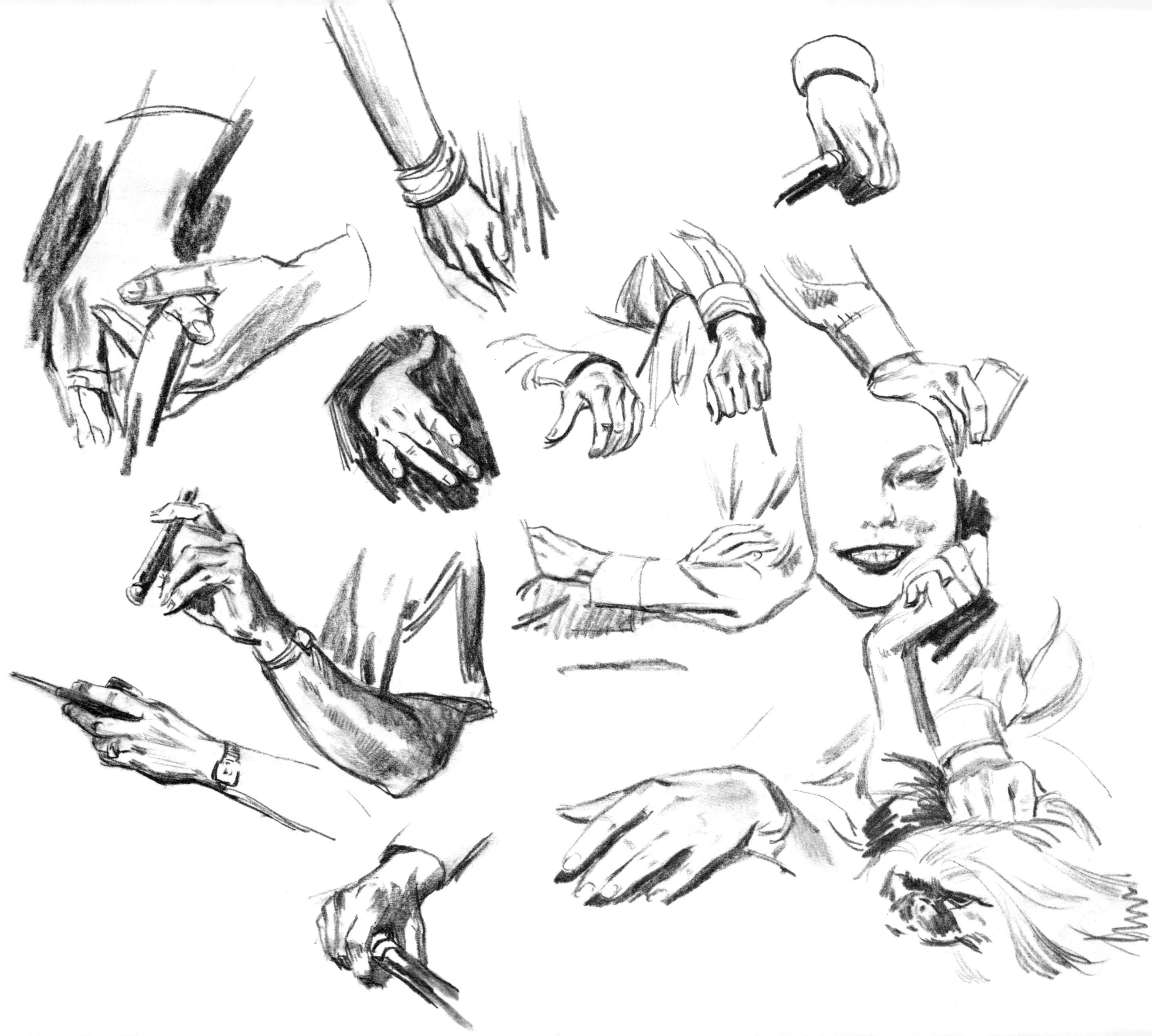

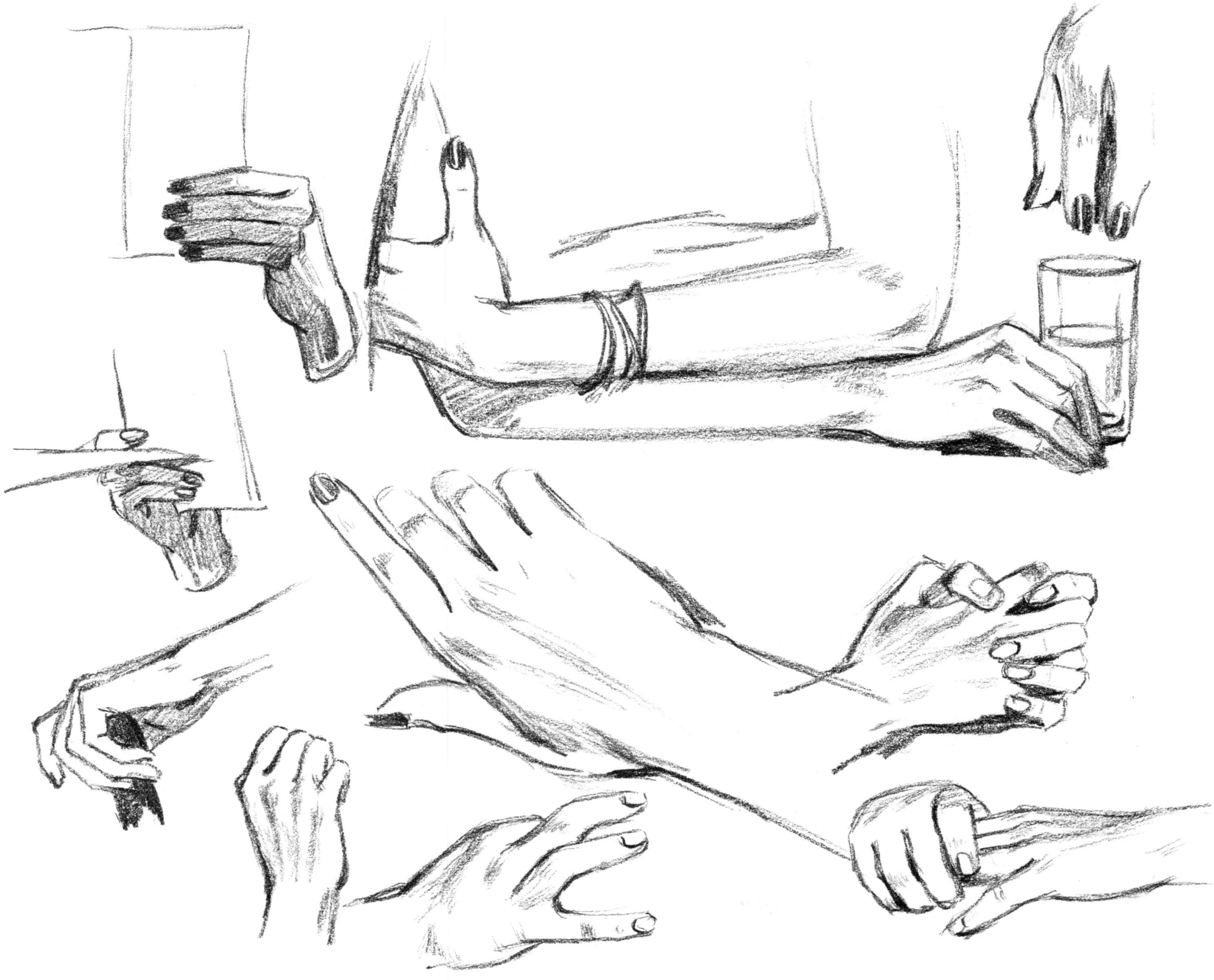

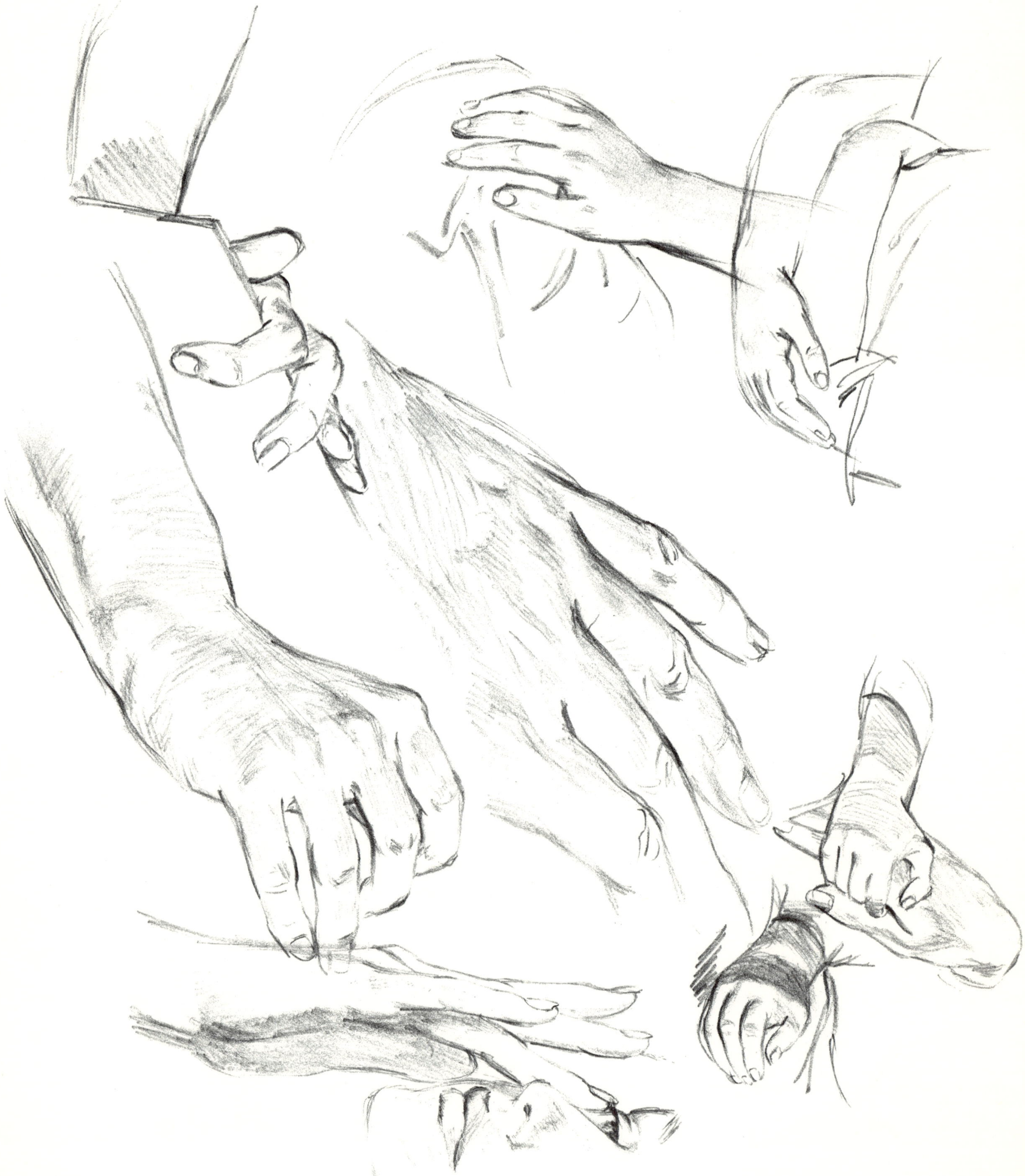

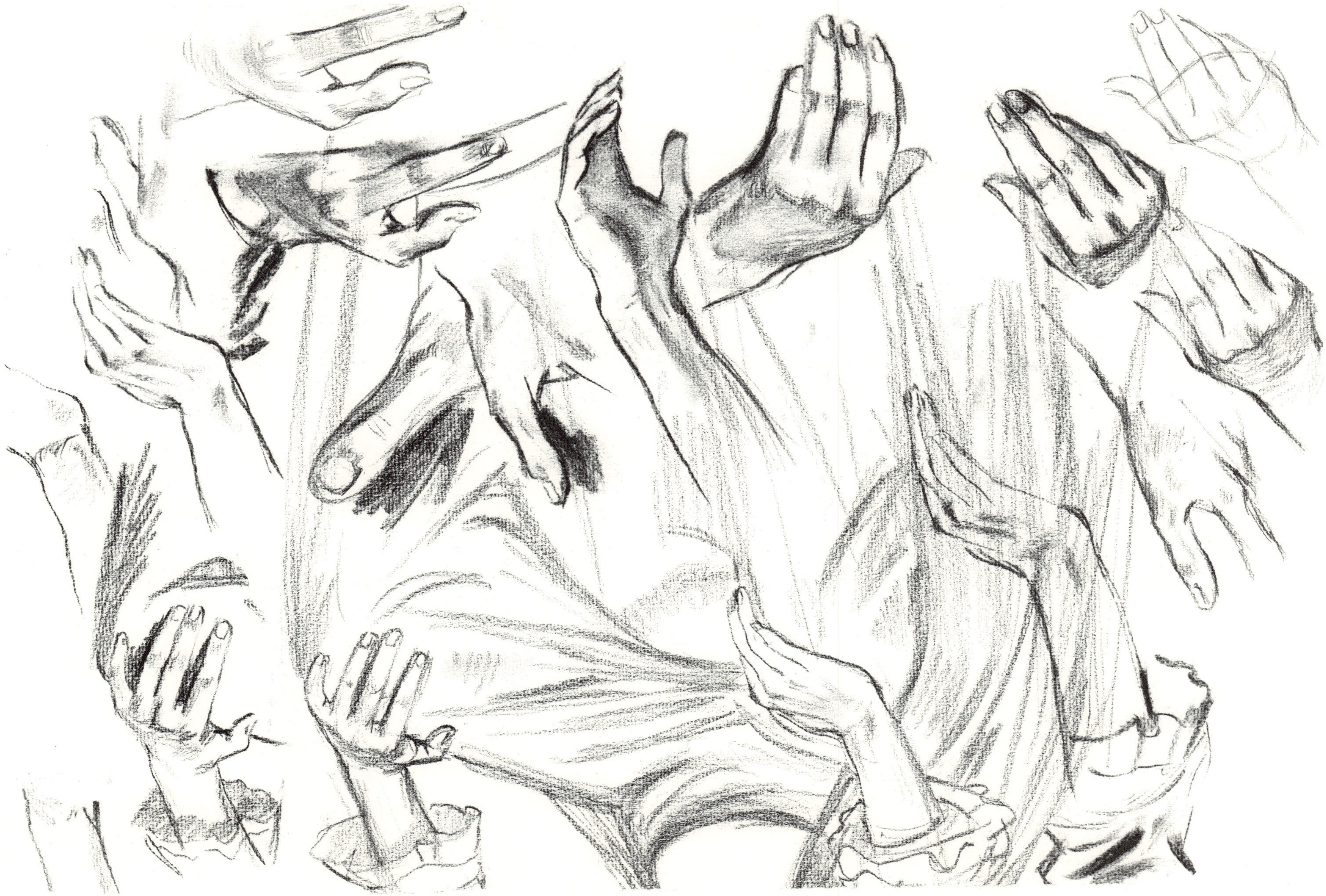

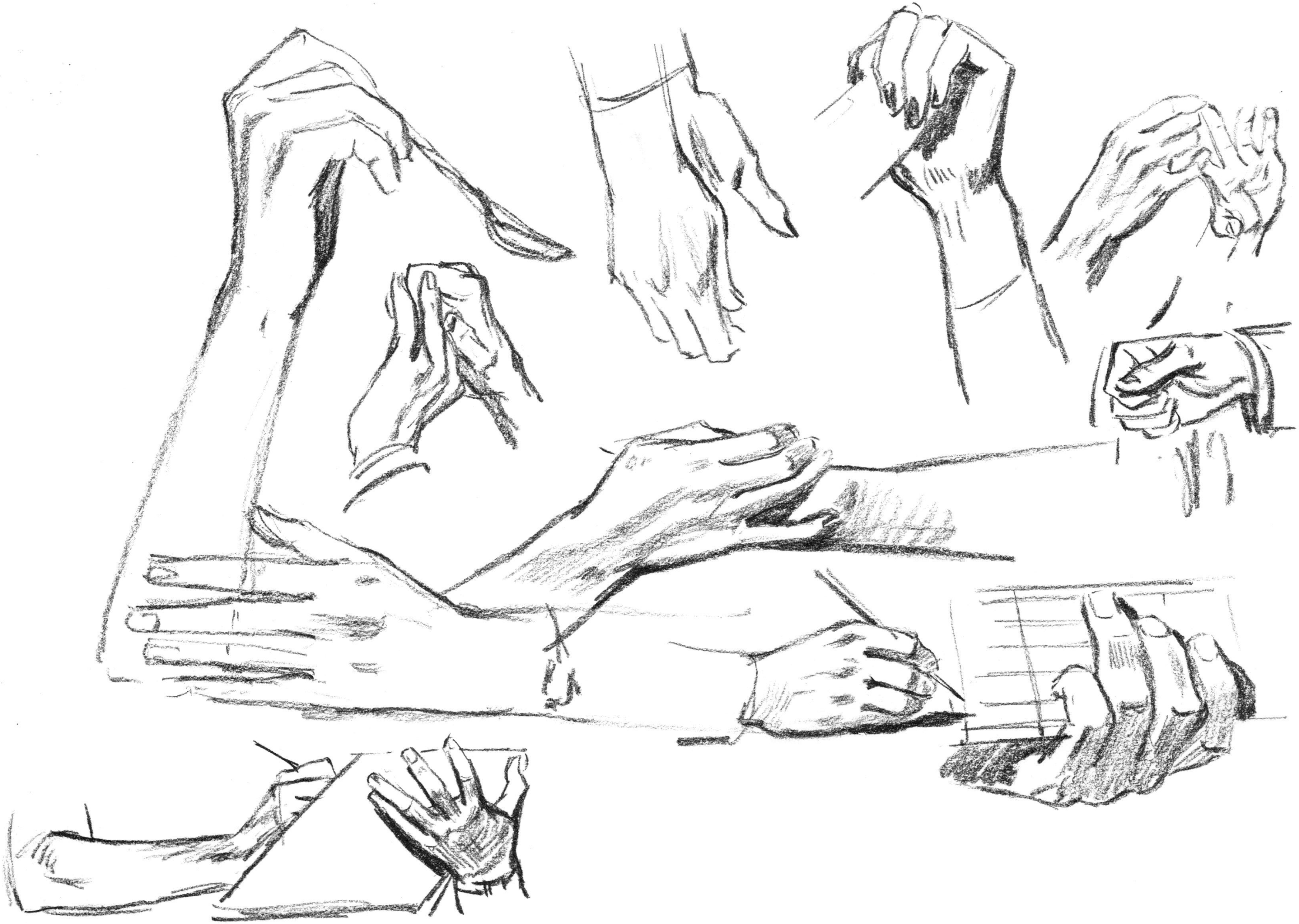

PORTFOLIO OF DRAWINGS OF THE OLD AND MODERN MASTERS

SOMETIME during the years between 30,000 B.C. and 12,000 B.C. man made his first attempts to record the shape of his hand. According to the findings of Professor Hugo Obermaier, a world authority on the drawings of Ice Age man, these were the first graphic efforts of that period, preceding man's drawings of the animals which cover the walls and ceilings of the Ice Age caves of France and Spain, (see *Art in the Ice Age* by Johannes Maringer and Hans-Georg Bandi, published by Fredrick A. Praeger, Inc. Library of Congress Catalogue card number: 52-13999).

By placing his hand or the hand of a friend against the dampened stone wall of the cave and then blowing a powder of iron oxide over it he left a negative silhouette or pattern of the hand on the wall. Two of the imprints show hands that have lost the two end phalanges of the fingers and the end phalange of the thumb. This could be, possibly, the first record of leprosy or ritualistic mutilation or maybe even the execution of early legal penalty.

The next recorded efforts to draw the hand, that remain to us, occur in the low relief decorations of the Assyrians and the paintings in the Egyptian tombs. Both the Assyrians and the Egyptians employed a stylistic approach in depicting the figure, namely, the head was shown in profile with the torso shown in front view and the legs again in profile. The hands and arms were always shown in actions that did not necessitate their being drawn in foreshortened views. The treatment of the hands was always a formalized formula, fingers were depicted as long and curvacious forms.

The Greeks were primarily devoted to sculpture in both the round and high relief. In the decoration of their magnificent vases and bowls they resorted to friezes of figures in various activities such as dancing, fighting and sports events coupled with running border designs of geometric pattern. In the drawings of the hands on these pieces of pottery the Greeks resorted to a stylized approach somewhat similar to that used by the Assyrians and the Egyptians.

In the mural paintings of India, done during the Gupta Period, 300 A.D., one finds drawings of hands that show considerable knowledge of physical structure and closely studied gesture. Hands are shown in foreshortened views and drawn most convincingly. There seem to be no actions of the hands that were avoided by their artists. Their painters took some liberties in drawing the figures, exaggerating the smallness of the waist, broadening the hips, enlarging the breasts and lengthening the eyes to conform with the Indian ideas of the ideal body. The hands, however, were drawn in normal proportion and with great grace. These paintings of the Gupta Period far surpassed in sophisticated draughtsmanship the works of artists working in Europe during the same era.

During the Dark Ages and the Middle Ages in Europe there was little or no effort by man to learn more about himself or the world in which he lived. The Church dominated his thinking and decreed that his thoughts should be devoted to the hereafter. It was more important to devote one's thinking toward the reward of Heaven or the torture of everlasting damnation than to spend one's efforts learning more about oneself and one's world. In this atmosphere the arts stagnated. This curious era continued for about 800 years.

With the revival of learning there came into being that age which we refer to as "The Renaissance". This well-known but loosely dated period was the strong hyphen between the Middle Ages and the era we call "Modern Civilization". It was the spawning ground for emancipated humanistic thought. It is not the province of this book nor the desire of the author to delve deeply into the history or reasons why the Renaissance came into being — however, there are a few significant advances in thinking and achievement that are worthy of mention. Paper was introduced into Europe, a Chinese invention, that replaced parchment as a common working surface for both the authors and the artists. Moveable type and the printing press came into being to replace the hand copied books made by monks in abbeys. The mariner's compass expanded navigation — gunpowder, another Chinese invention, revolutionized the "art" of war. The Ptolemaic concept of astronomy was replaced by the Copernican theory. The writings of Dante, Petrarch and Boccaccio were original, monumental, works of art. It was in fact the works of these authors that did most to free the graphic artist of the day.

In this rich soil of intellectual inquiry all the arts flourished, none more so than the art of painting and drawing. By some peculiar fortuitous circumstance there appeared, virtually simultaneously, a group of artistic giants, Michelangelo, da Vinci, Raphael, Titian, Bellini, del Sarto, Giorgione and on and on. The geometry of sight, which we call perspective, was discovered and perfected. The popes and cardinals of the church, coming from the ruling houses of Rome and Florence, were champions of the new humanist thought. Artists could now study the anatomy of man without fear of suffering the wrath of the Church. Men began to draw from the posture of knowledge instead of simple visual observation of surface. Drawings showed a new vigorous authority. The seed of man's inquisitiveness broke into bloom during this period and bore the fruit that produced Modern Man. The author suggests that those students who have not been exposed to the history of the Renaissance would profit greatly by devoting time to its study.

WASHINGTON ALLSTON 1779-1843

Allston was born in South Carolina in 1779, was educated at Harvard and spent most of his life in Boston. He traveled abroad during 1803 and 1806 to London and Paris. Living two years in London and a year in Paris, he attempted to broaden his education and training as an artist. From Paris, in 1806, he went to Italy, where he became enamored of Venetian painting. Titian, Veronese and Tintoretto were among those he especially admired. Before his return to America, Allston's paintings were large, subjective canvases, romantic visions of nature and life. Through it all, however, the influence of classical art can be observed. In his work, Allston reveals the awe-inspiring grandeur of nature, and he was one of the first American artists to be so influenced. It was after 1818 that his work began to reveal a lyrical dreamlike quality. Also, at this time, his work took on a much smaller scale. Allston, it can be said, set new standards for American painting, and thus he stands out in the history of its development.

It will be noted that Allston in his development of the form in this drawing used modeling that went across the form. Note the strokes of the crayon at (1) which give the feeling of the volume of the form.

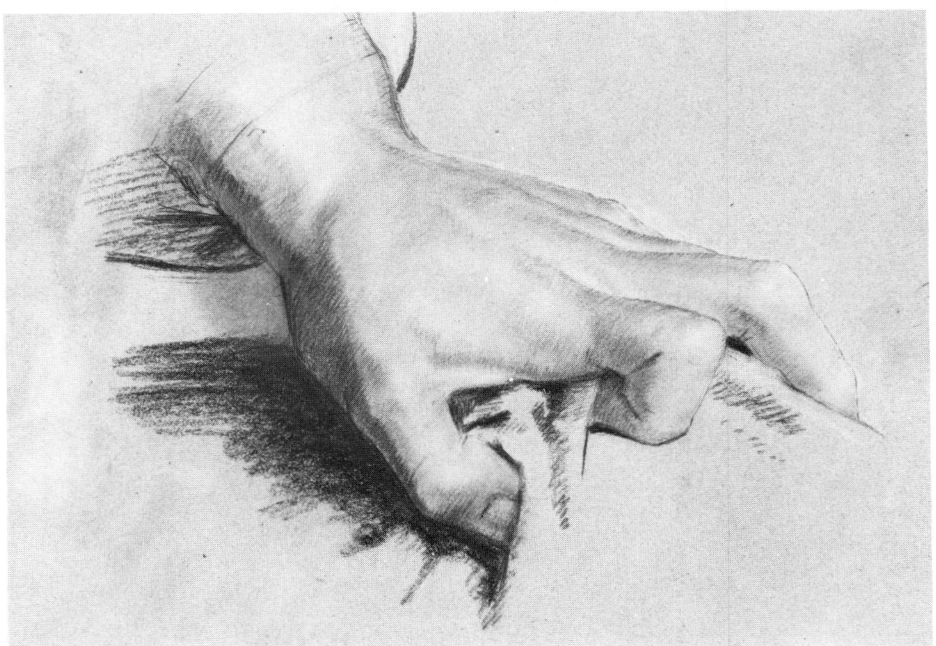

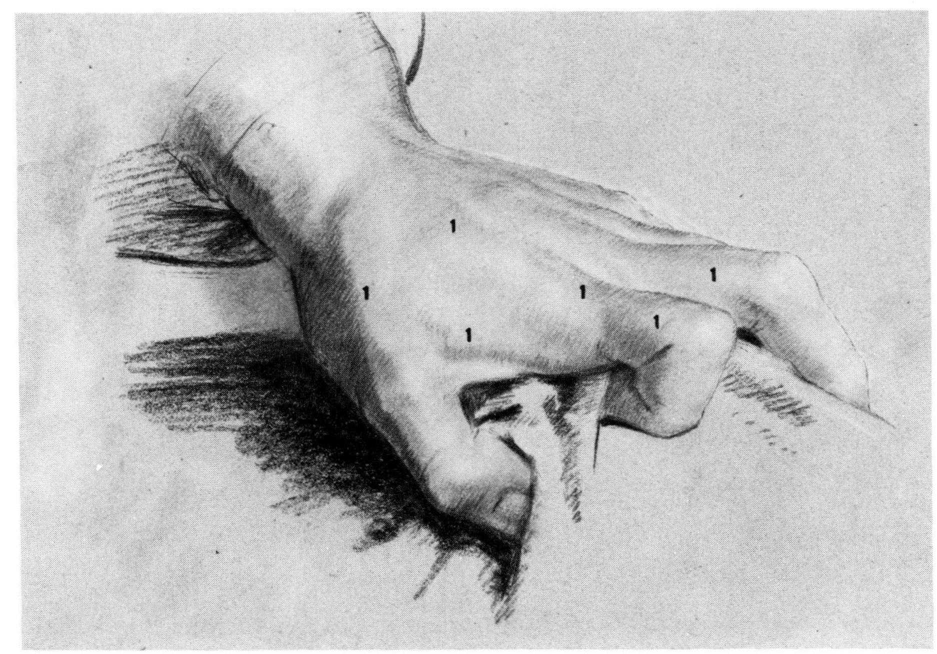

Charcoal heightened with White Chalk
FOGG ART MUSEUM, HARVARD UNIVERSITY, CAMBRIDGE, MASS.
WASHINGTON ALLSTON TRUST

SIR LAURENCE ALMA-TADEMA (1836-1912)

Alma-Tadema was born in the Netherlands. He entered the Antwerp Academy at the age of sixteen and studied under Gustav Wappers and Baron Henri Leys. He was a precocious student and became Baron Leys' assistant in painting the murals in the Antwerp City Hall in 1859. He won the Gold Medal for painting in Amsterdam at the age of twenty six. His most characteristic work was in the area of scenes from Greek and Roman subject matter. In 1870 he went to live in England, where he enjoyed great success. In 1876 he was elected associate of the Royal Academecian. Queen Victoria knighted him in 1899. His works were remarkable for the realistic rendering of naturalistic textures. He died in London in June, 1912.

In these drawings Alma-Tadema has contented himself with naturalistic value pattern in drawing No. 1. Both of these drawings were made to be used as models to paint from. One can see that drawing No. 2 is labeled "Marc Antony", obviously intended for a painting dealing with Roman subject matter. In drawing No. 2, great care has been taken to record the outline of the hand and fingers. The modeling in the back of the hand has been subtly drawn to indicate tendons of the fingers (3) and the first interosseus (4).

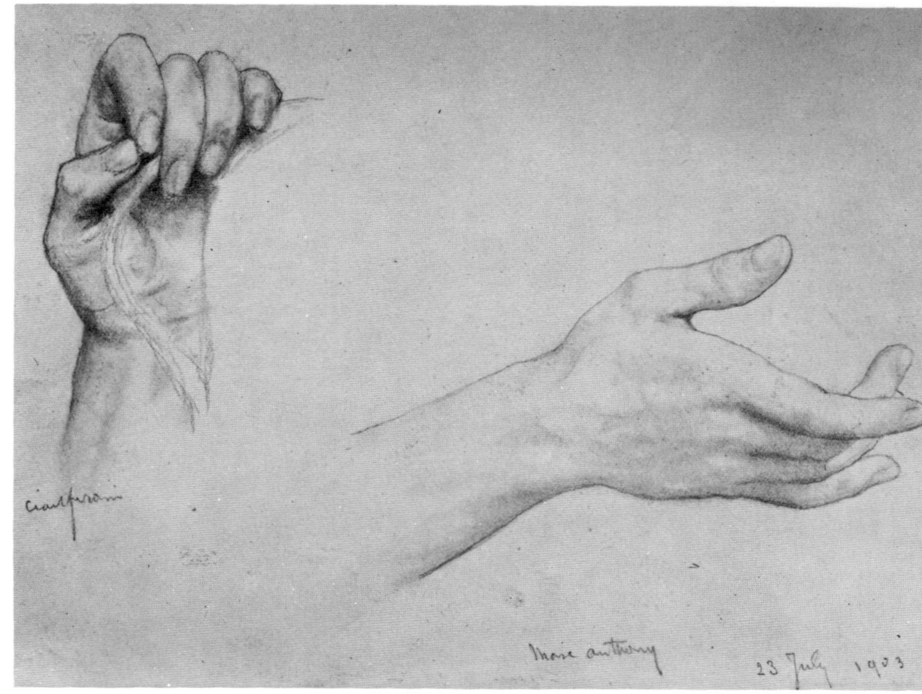

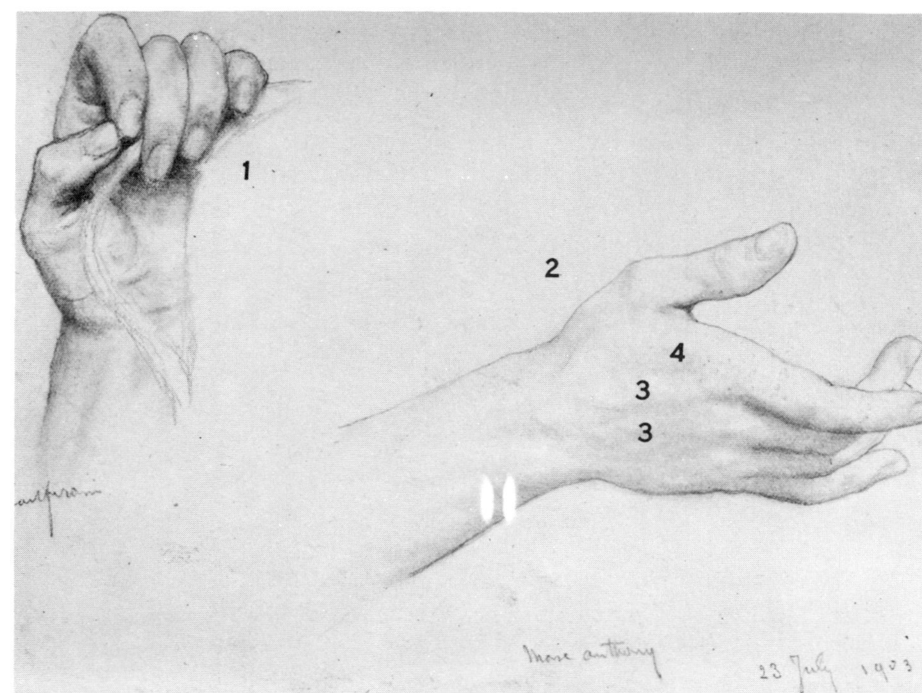

Pencil

THE BRITISH MUSEUM, LONDON, ENGLAND

ANTOINE WATTEAU (1684-1721)

This charming French painter was born to humble Flemish parents. At fourteen years of age he became the apprentice of a mediocre painter, Gérin of Valenciennes. When Gérin died, Watteau was left penniless; he went to Paris to work for a scene-painter, Métayer. Watteau, whose health was never robust, worked in a factory that turned out religious paintings wholesale. For this he was paid three francs a week and meager food. In 1709 he won second place in the Prix de Rome competition. Watteau was elected associate of the Academy in 1712 and full member in 1717. His diploma painting, *The Embarkment for Cythera* now hangs in the Louvre. Watteau was invited to live in the household of Crozat, the greatest private art collector of his time. While a guest of Crozat, Watteau painted four decorative panels of *The Seasons* for his benefactor. For six months he lived with his friend Gersaint, an art dealer. His health forced him to live in the country. Realizing that the end of his life was near at hand, he worked feverishly and continuously. His last work was a portrait of the Italian pastellist Rosalba Carriera, who at the same time painted Watteau's portrait. On July 18, 1721 he succumbed to the tuberculosis that had pestered him since youth. He died in the arms of his friend Gersaint.

The grace and charm of these drawings is a delight. Watteau was an accomplished draughtsman who drew with great clarity. He was particularly successful in establishing the mood of a pose. Note the feminine character of each of these drawings. He drew directly with no fumbling with either line or tone. Watteau's drawing of drapery of costumes is particularly worthy of study by the student.

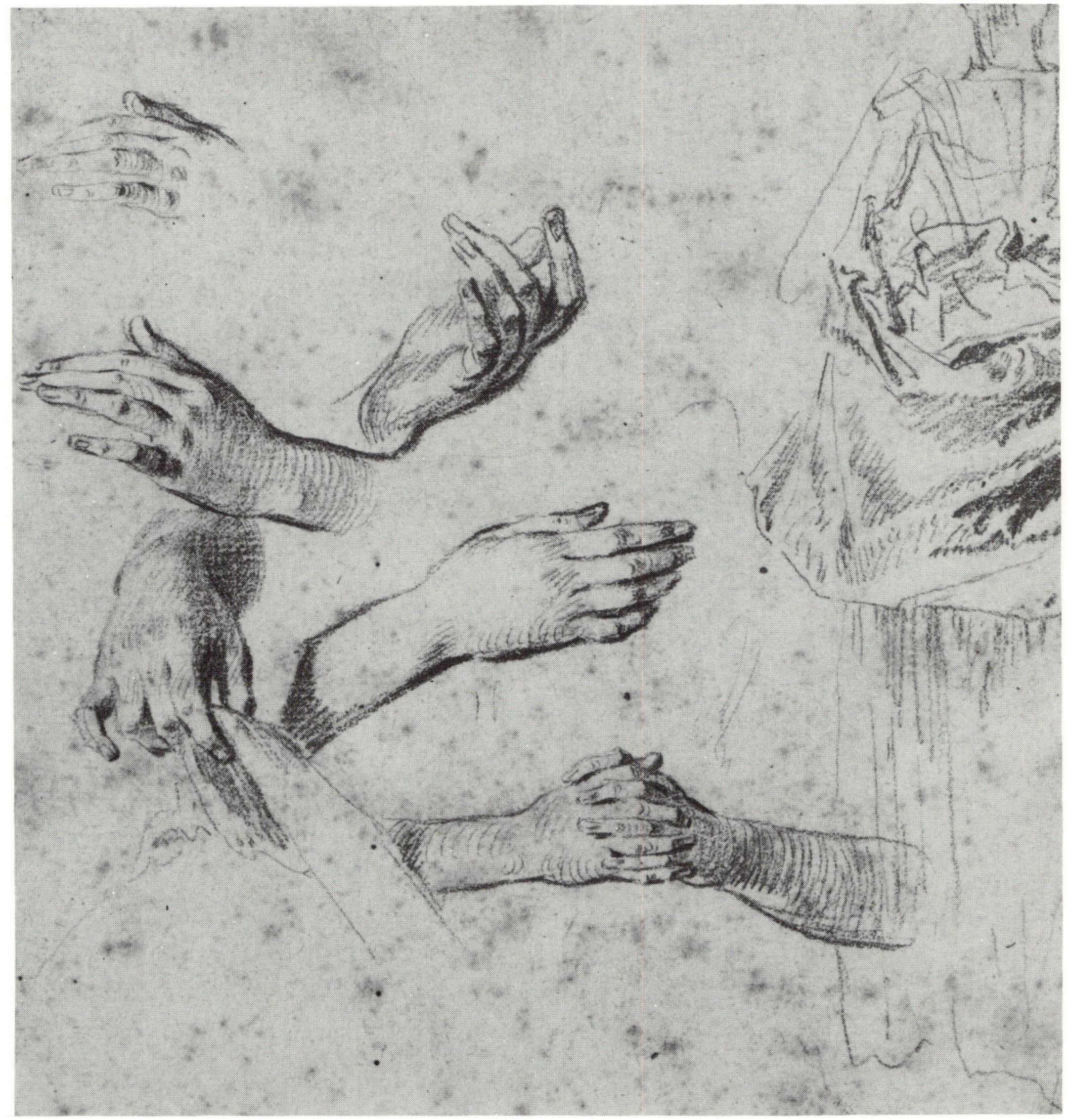

Red Chalk
THE BRITISH MUSEUM, LONDON, ENGLAND

SIR DAVID WILKIE, R.A. (1785-1841)

This Scottish painter was the son of a parish minister. At the age of fourteen he began his study at the Trustees' Academy in Edinburgh. His early work was influenced by the paintings of Carse and David Allan who painted scenes of common life. Wilkie haunted fairs and public markets filling countless sketch-books with drawings that he later used in his paintings. In 1805 he went to London where he entered the Royal Academy school. In 1809, at twenty-four years of age, he was elected associate Royal Academician and two years later was elevated to Royal Academician. In 1830 he succeeded the painter Lawrence as painter to the king. From that period forward he painted portraits of royalty and distinguished personages. Wilkie never flattered his subjects and his portraits were seldom accepted with satisfaction. Wilkie's popularity was due to his genre painting and his study of the "Little Masters" of Dutch painting influenced his works. Most of the pictures in this category are early works and are noted for their detailed handling and sureness of touch. Wilkie went to Spain in 1825 and is reputed to be the first English painter to visit that country. He died and was buried at sea. Turner was moved to paint a commemorative work, *Burial at Sea of Sir David Wilkie* that now hangs in the Tate Gallery in London. Wilkie ranks with Lawrence and Bonnington as a romantic painter. Due to the vagaries of fate he has never achieved greatness.

This drawing of a woman's hand is full of the character of a painter's technique in drawing. The value pattern seems to be his major concern until one studies closer the subtle drawing of the flesh pads of the palm and the fingers. There is great accuracy in the drawing of the area of the wrist.

Black Chalk

THE BRITISH MUSEUM, LONDON, ENGLAND

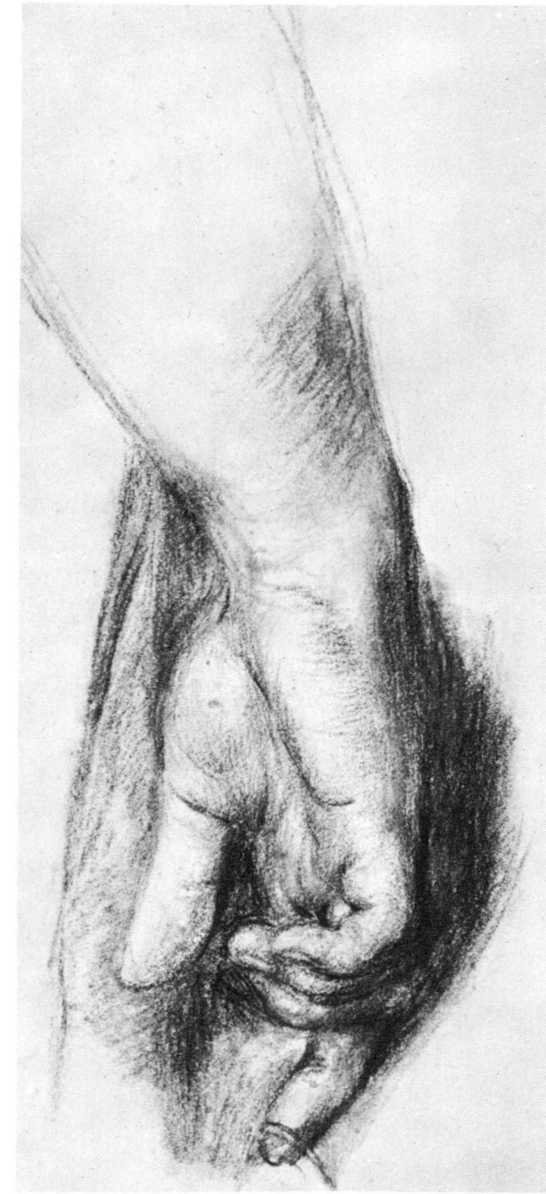

FRANCESCO CAVALIERE VANNI (1565-1609)

Francesco Vanni was born in Siena. His step-father, Arcangiolo Salimbene, taught him the rudiments of design. Salimbene died during Vanni's early youth. Upon reaching the age of sixteen Francesco left Siena and went to Bologna. There is no record of what he did in Bologna, however, we do know that he went to Rome and entered the school of Giovanni de Veschi. He also studied under Federigo Barocci who was a well known painter of religious subjects. Vanni imitated Barocci's style with considerable success and was soon receiving commissions for pictures for churches. He returned to Siena where he worked for a relatively short while before moving to Parma. He later went to Bologna again and then went on to Rome. In Rome he painted a "Simon Magus" for the church of Saint Peter. He was successful in getting additional commissions to paint pictures for a number of churches in Rome; among the most important ones are "Saint Michael vanquishing the rebel Angels", in the sacristy of the church of Saint Gregorio; a "Pieta" in Santa Maria in Vallicella; and the "Assumption", in Saint Lorenzo in Miranda. At Siena, in the chapel of El Refugio there is a painting of Saint Catherine and a "Saint Raimondo walking on the Sea" in the church of the Dominicans. Vanni died in Siena in 1609.

Sanquine Crayon

HERMITAGE MUSEUM, STALINGRAD, RUSSIA

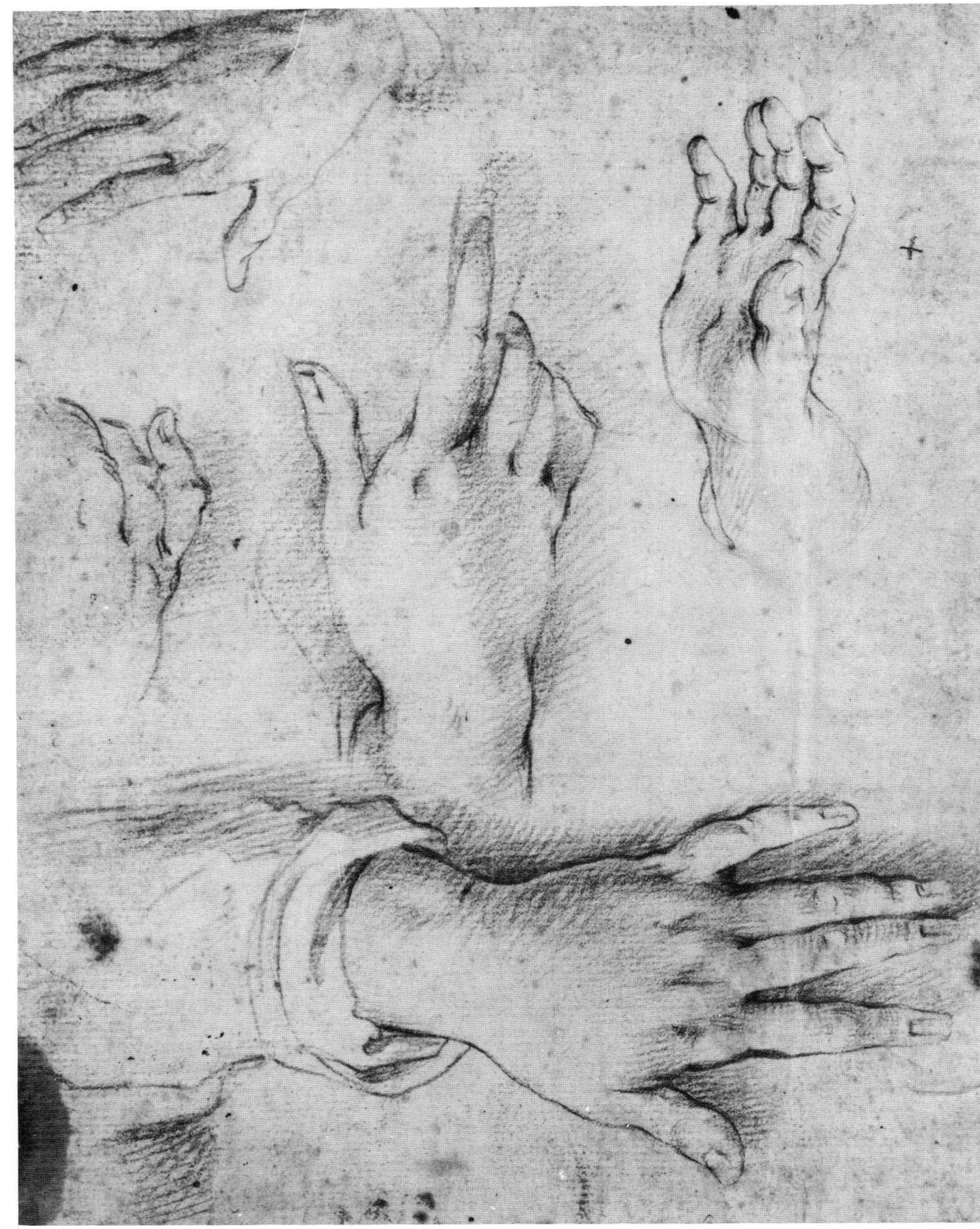

KATHE KOLLWITZ (1867-1945)

German etcher, lithographer and sculptor. Kollwitz's husband was a physician who worked among the poor in Germany and she accompanied him on many of his rounds. This close contact with the underprivileged was admirably recorded by her in numerous drawings and etchings of the miseries suffered by those peoples. *Monument to the Dead* in the military cemetery in Essen, Belgium — a memorial to her son who was killed in battle in 1914 — is a work of great power and emotional impact. She left many self-portraits that are marvels of draughtmanship. She was persecuted by the Nazis during the World War years.

These drawings in pen and ink of the artist's left hand are superb examples of using a medium that does not lend itself easily to producing tonal delicacies but here controlled magnificently (1). The form of the fingers and the knuckles has been executed with great care and technical charm. The technique is obviously one that is related to etching (2). The shadow of the arm and hand were painted in with a brush (3). A truly superb pair of drawings.

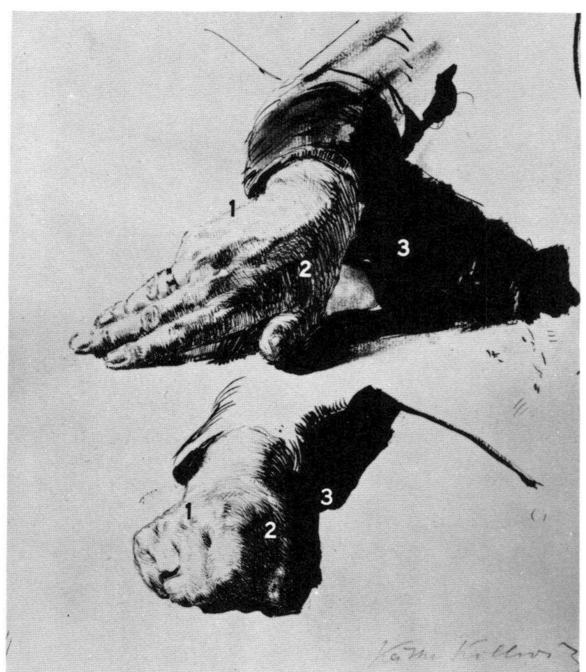

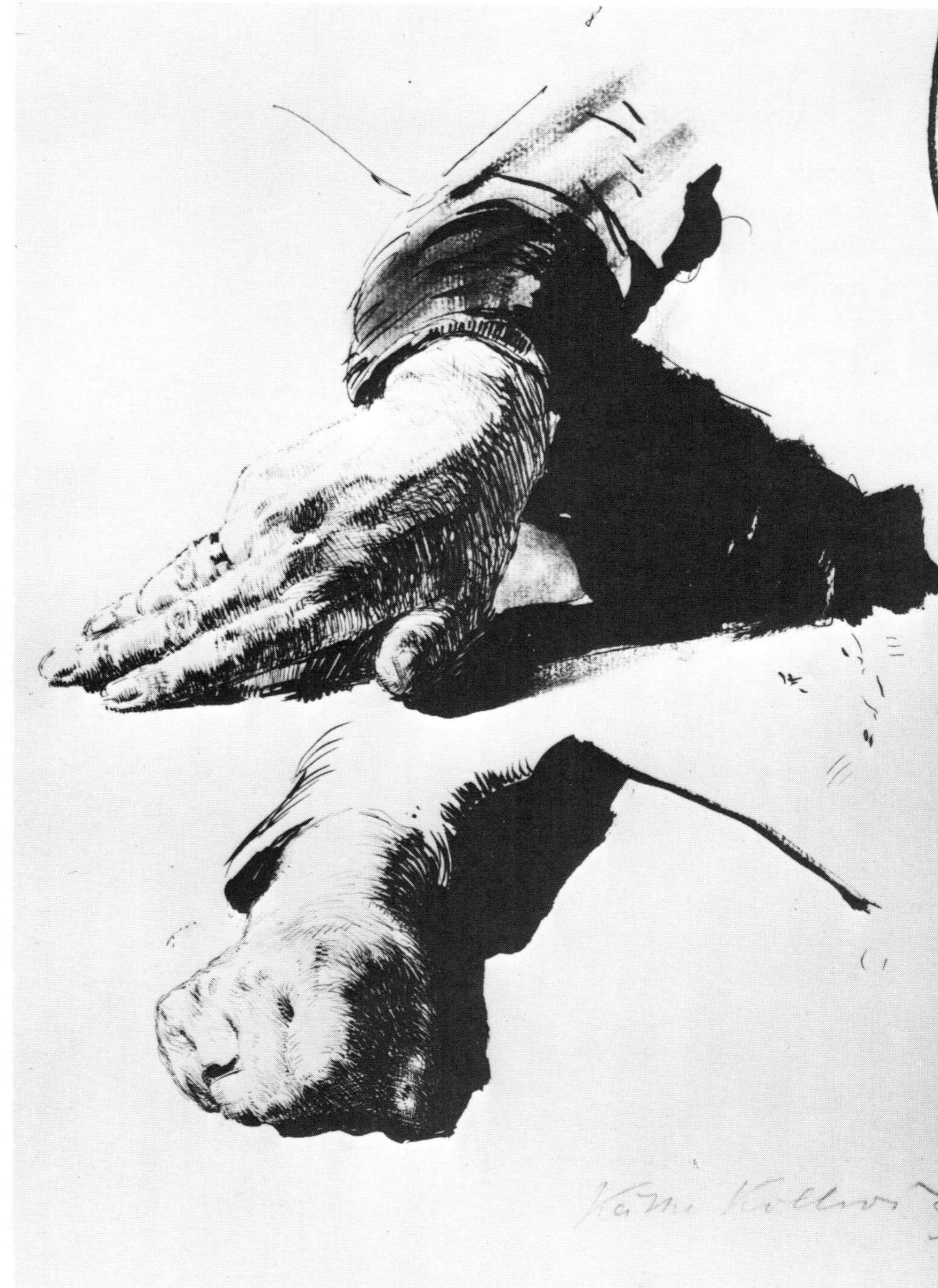

Pen and Ink
THE BRITISH MUSEUM, LONDON, ENGLAND

WILHELM LEIBL (1844-1900)

German painter of peasant and bourgeois subjects. Born in Cologne, Germany, Leibl was influenced by the works of the Dutch masters in the early part of his career and the paintings he made during that period are masterpieces of Naturalism. Later in his career he turned to a more open technique, influenced by Impressionism. He was a very strong draughtsman and left a large body of work in this area. He enjoyed, almost from the start, both public and critical acclaim. Living with Gustave Corbet for nine months in Paris had a tremendous influence on Leibl's technique. His attachment to Realism was greatly strengthened by this association. In 1870 he was awarded the Paris Gold Medal for his *Portrait of Frau Gedon,* which clearly showed the influence of Rembrandt. The perfect draughtsmanship he exhibited through his attention to every small detail, gained for him his influence on the New Objectivists of the 1920's, particularly in figure studies and portrait drawing.

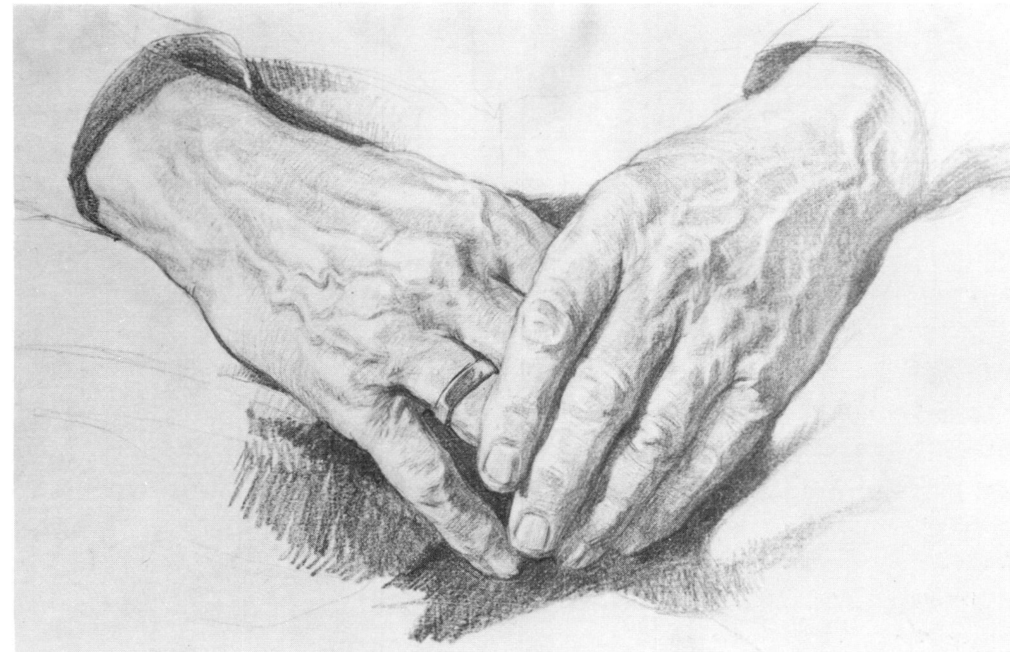

This drawing done with charcoal pencil is a good example of his care in recording surface detail. The veins of the hand (1) have been closely observed. The direction of the lines (2) have been drawn across the form to indicate the volume of the hand. The detail of the knuckles (3) has caught the character of the time-worn hands. The observation of the reflected light has added to the conviction of the form of the fingers. This drawing was obviously made as a study from which to paint.

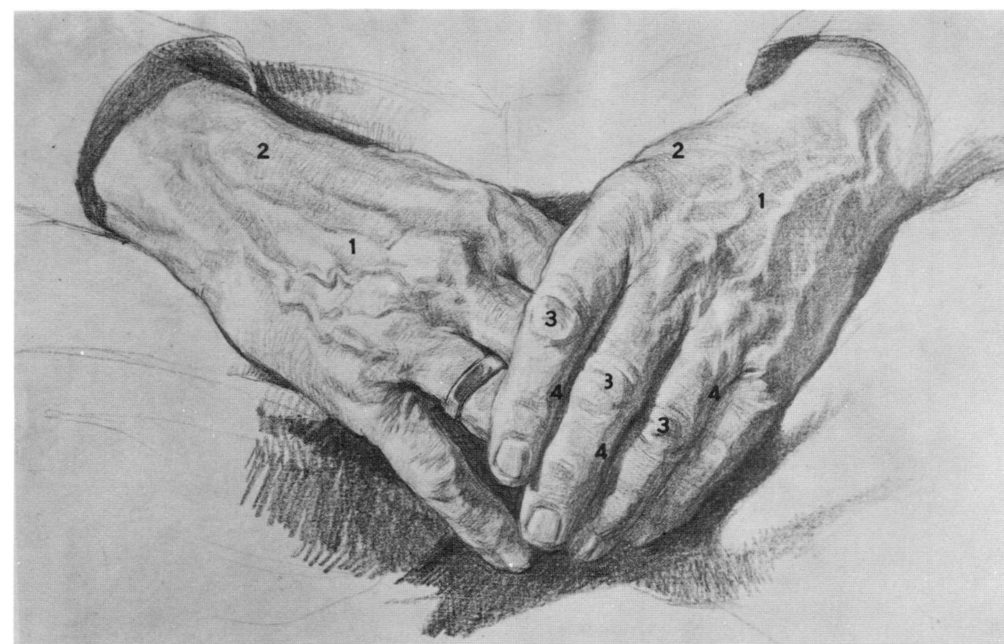

Charcoal Pencil

VICTORIA AND ALBERT MUSEUM, LONDON, ENGLAND

LEON KROLL (1884-)

Leon Kroll, Americain painter and lithographer born in New York City, studied at the Art Students' League and the National Academy in New York. He also studied abroad in Paris with Jan-Paul Laurens. Kroll's early works were strongly painted urban and industrial subjects — somewhat related to the style of the New York realists. His later paintings are basically academic in style reflecting in great part the influence of Laurens. His subject matter shows his preference for still-life and figure studies. Kroll has painted murals for the Justice Building in Washington, D.C. (1937) and the War Memorial in Worcester, Massachusetts (1941). He gained much inspiration from his friendship with George Bellows, whose untimely death had a tremendous saddening effect on Kroll as well as the other artists of his day.

This drawing in black crayon shows Kroll's interest in establishing the pattern of light and shade on the fleshy pads of the hand. The accented line (1) is that of a painter. This drawing was made with the purpose of using it as reference from which to paint. Note the broken line on the drawing of the index finger (2) a further indication that this drawing was done by a painter.

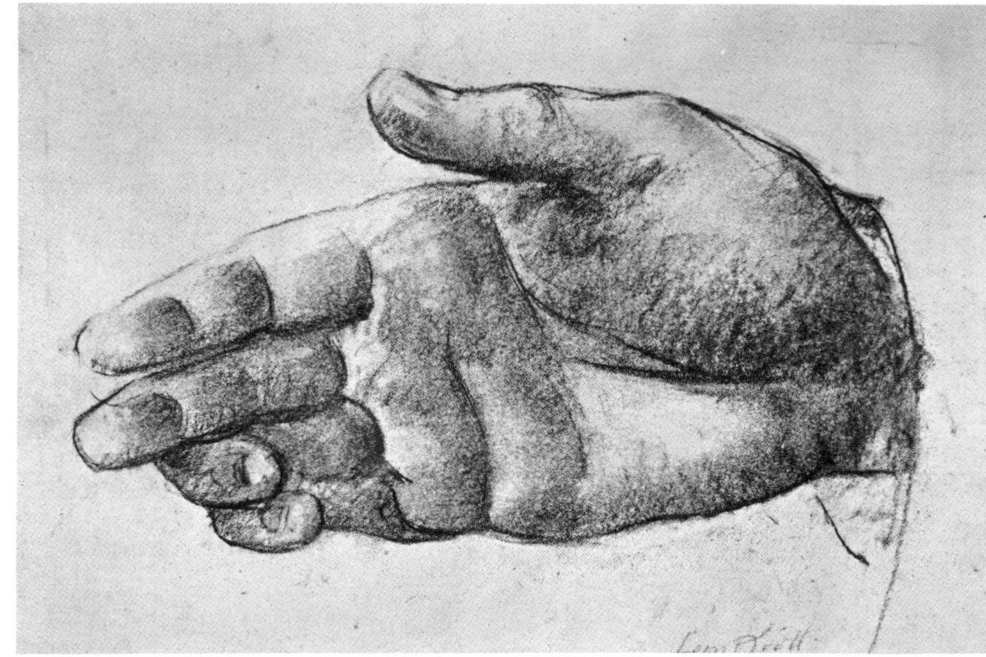

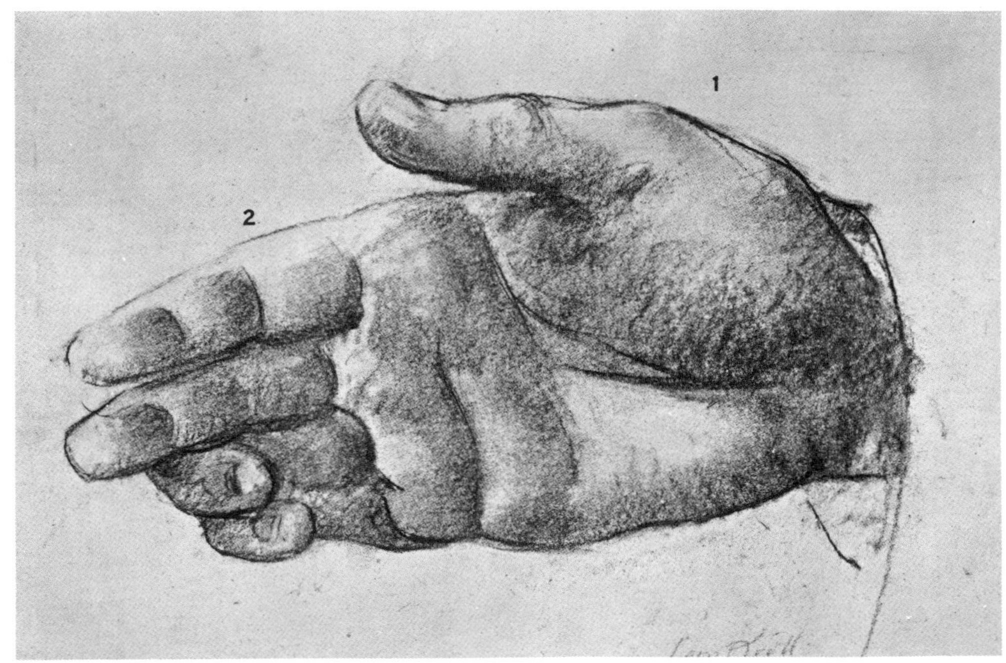

Black Crayon
LEON KROLL COLLECTION, NEW YORK CITY, NEW YORK

BARON FREDRICK LEIGHTON (1830-1896)

English classical painter of heroic subjects, Leighton's father was a physician of some wealth, which permitted the son to study in Rome, Dresden, Berlin, Frankfurt, Brussels and Paris. Leighton painted in the grand traditional manner of European painting. His works show his prim-highminded-idealistic-unemotional approach to painting. He was an exacting draughtsman of naturalistic effects. He enjoyed huge success in Victorian England where his idealistic works met with the demands of both public and critics.

These pencil and wash drawings show the care with which Leighton studied his subject matter. (1) In the originals of these drawings the delicacy of tonality (2) is quite delightful. The accented line (3) is handled with great care to state the volume of the form. The value pattern has been done with exact attention to light intensity (4).

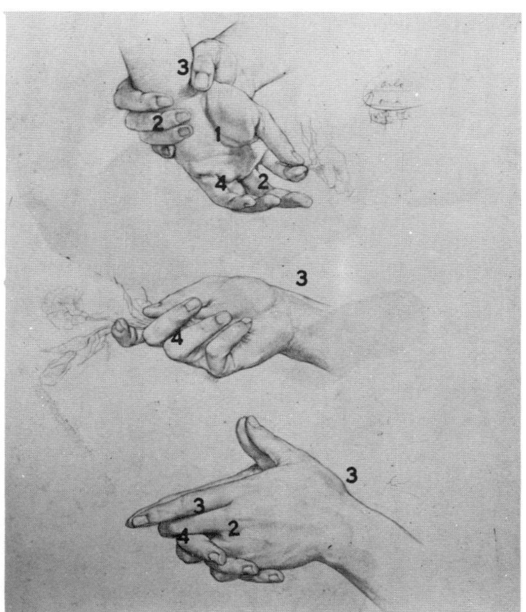

Pencil and Wash

THE BRITISH MUSEUM, LONDON, ENGLAND

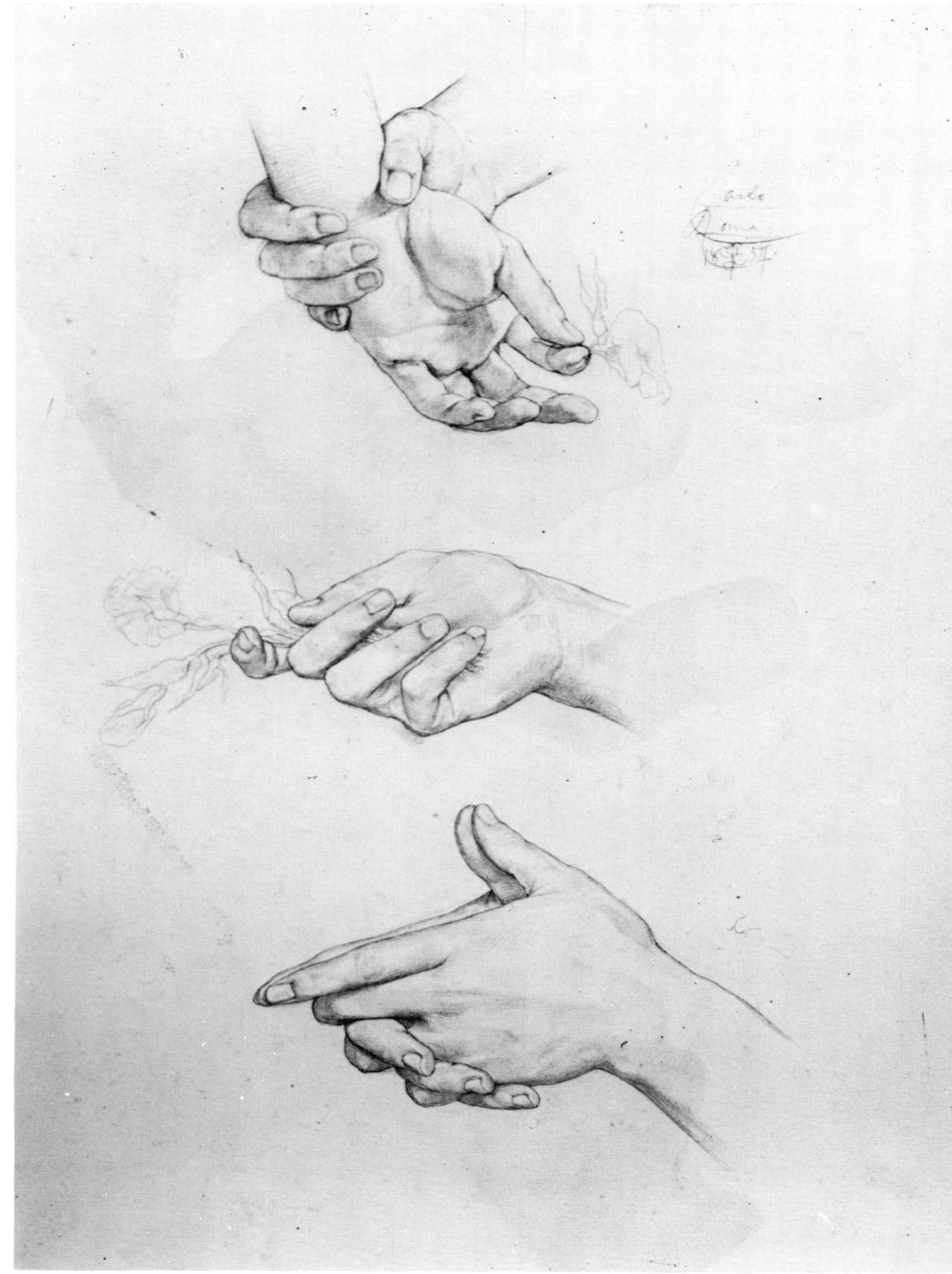

FRA FILIPPO (1406-1469)

Fra Filippo Lippi was one of the most outstanding early Renaissance painters. He was influenced in his early work by the paintings of Masaccio. He was a Carmelite monk until 1431 when he left the order to work as an independent painter. He was the chaplain of a convent and abducted one of the nuns who bore him two children. He was always tardy in his commissions and was on one occasion convicted of forgery. He had friends in the powerful Medici family and he was pardoned. One of his sons became a monk and a great painter.

The distinguishing quality of these drawings is their graceful, fluid quality. Though the drawing of the structure at (1) in the back of the upper hand is weak and flat, the gesture of the fingers and their structure is convincing. The structure in the second hand (2) is stronger in its drawing and the modeling at the wrist is well stated.

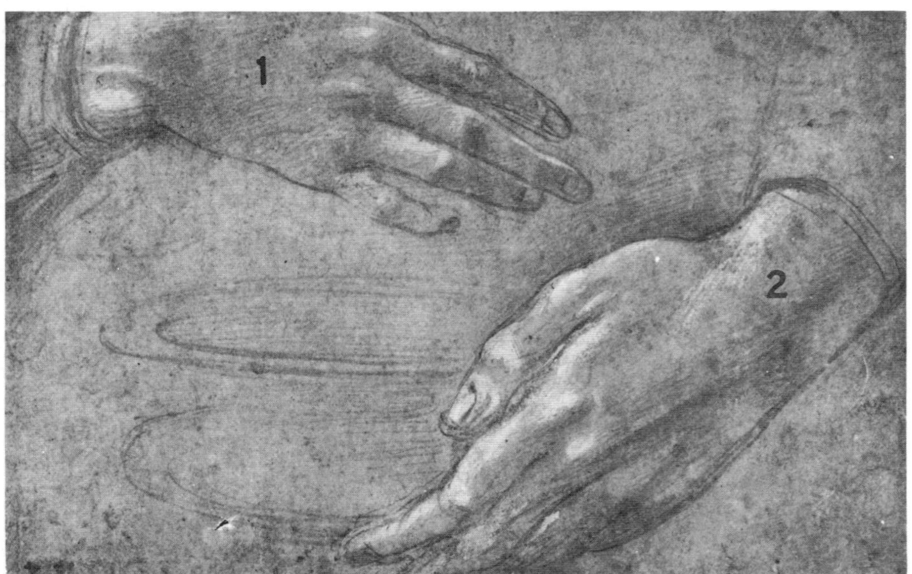

Black Chalk, heightened with White
THE BRITISH MUSEUM, LONDON, ENGLAND

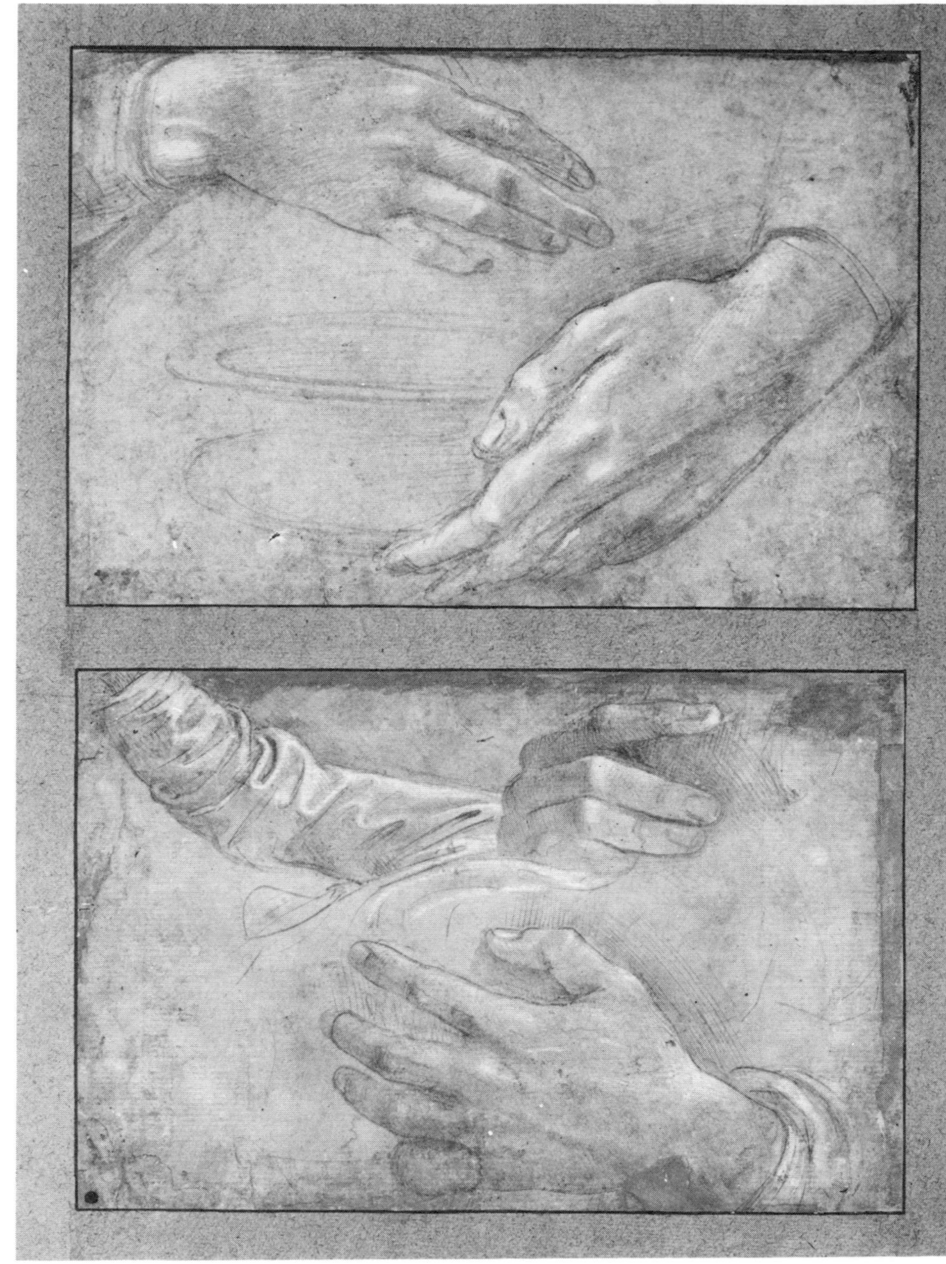

ALBRECHT DÜRER (1471-1528)

Dürer was born in Nuremberg, Germany; his father was a goldsmith. Dürer was the second of eighteen children and the favorite of his father who apprenticed him to the leading artist of Nuremberg, Michael Wolgemut, a worker in woodcuts. The great mass of Dürer's work was in the area of woodcuts and copper engraving. He was one of our earliest illustrators. The printing press made it possible to spread abroad the engravings of Biblical subjects that masters, such as Dürer, executed. In 1490 Dürer traveled to Italy to see the works of the masters who were the beginning of the Renaissance in painting.

This drawing is one of the best known drawings of hands in the world. Drawn with a quill on blue paper, it shows the training of Dürer as an engraver. Note the "open" strokes of the pen in white ink at (1) and the cross-hatch at (2) to indicate the form.

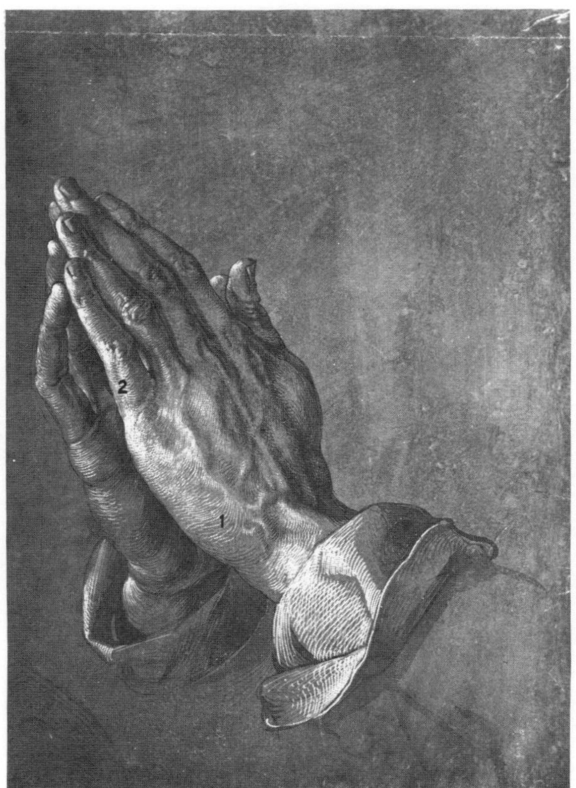

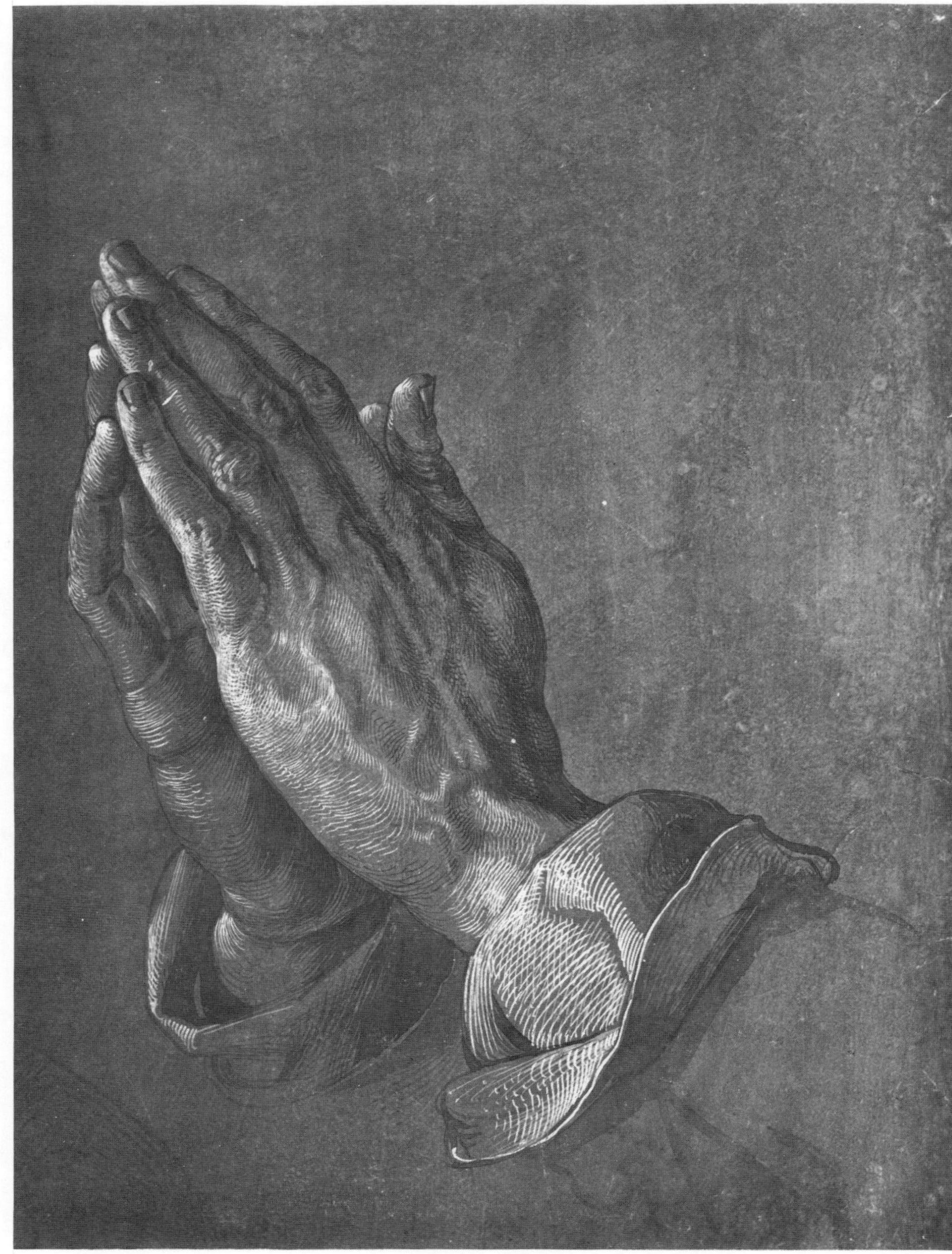

Pen and Ink on Toned Paper heightened with White
ALBERTINA MUSEUM, VIENNA, AUSTRIA

ALBRECHT DÜRER (Continued)

With the exception of Raphael and Rubens, few artists have enjoyed as successful a career as Dürer's. His fame spread all over Europe and his friendship with the great painters of the day and the crowned heads is legendary. Statesmen, humanists and the great reformers of the day all paid him homage. Raphael felt himself honored by an exchange of drawings with Dürer. His life seemed to be totally free of the struggles, jealousies, and annoying events that plagued many other artists. The one unhappy note in his life, if we can believe historical record, was his wife who was reputed to be a terrible shrew and money-pincher. For years the name of Agnes Dürer was used as an example, throughout Germany and the rest of Europe, as the unworthy wife of a great man. Dürer was never known to have referred to her with any affection or tenderness, saying of her that she was his house mate and helper in the selling of his prints. However, in her behalf it is said that following Dürer's death, Agnes Dürer showed generosity to his brothers. The enormous quantity of engravings, executed by Dürer, forces one to say that he was primarily a woodcutter and an engraver instead of a painter. However, he did leave for posterity a great number of paintings that are still considered as masterpieces of his age.

In this drawing one can see again Dürer's use of a line that is generally associated with preparation for engraving (1). At (2) note that the linear modeling is across the form to increase the feeling of volume. Dürer seldom used a dry, hard medium, such as chalk with which to draw, preferring to use the open line of the quill to achieve his modeling of form.

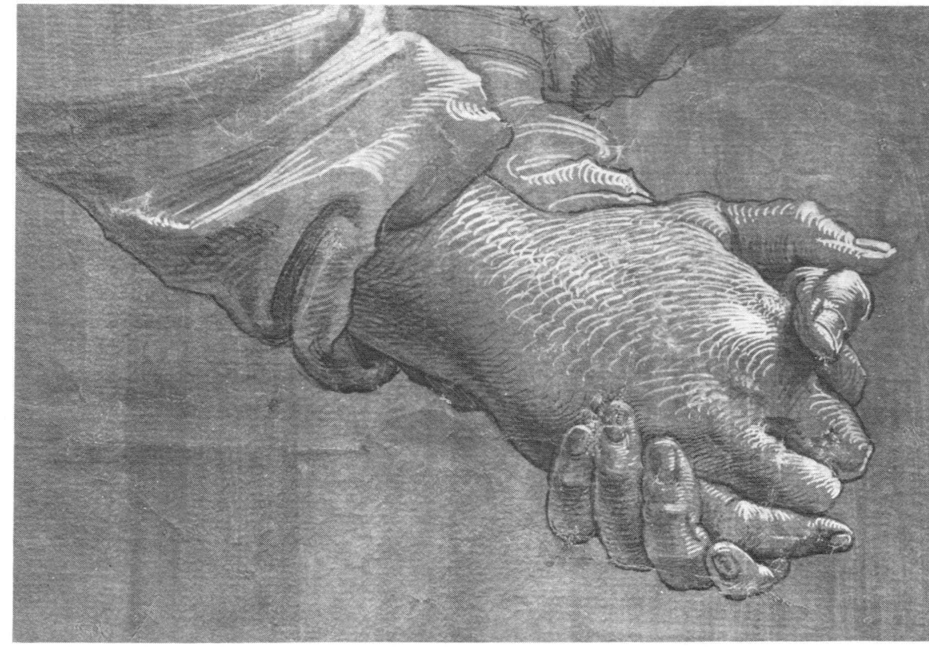

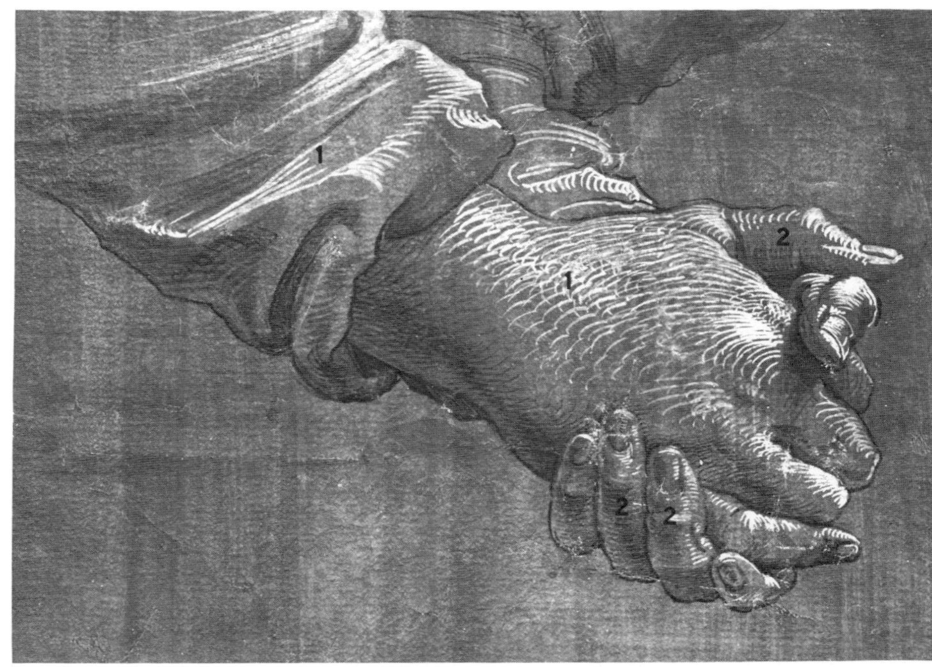

Pen and Ink on Toned Paper heightened with White
ALBERTINA MUSEUM, VIENNA, AUSTRIA

BARTOLOMEO BANDINELLI (1493-1560)

Born in Florence where his father was a goldsmith for the Medici family, Bandinelli was an arrogant and egotistical student who was much disliked by his fellow students. He was a trouble maker. His great desire was to surpass Michelangelo; he never even came close. His most colossal work was a choir screen for the Duomo in Florence. It comprised eighty statues and was an artistic failure and was removed from the Duomo. He was primarily a sculptor; he left many very good busts and hundreds of drawings.

These drawings of hands are drawn with great strength and show the sculptor's interest in form. Both drawings demonstrate the artist's solid knowledge of anatomical structure. Note the geometric quality of the drawing of the wrist at (1). Note the strong drawing of the tendons over the knuckles at (2). The quality of line (4) demonstrates the assurance of Bandinelli's knowledge — it is authoritative.

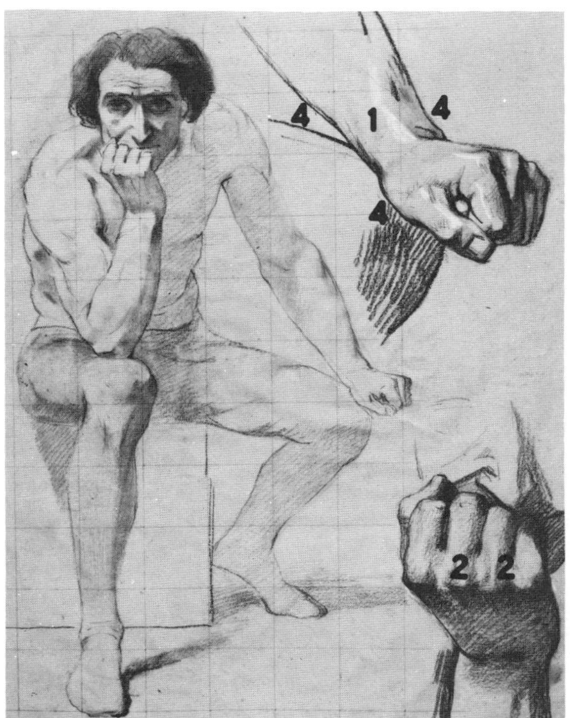

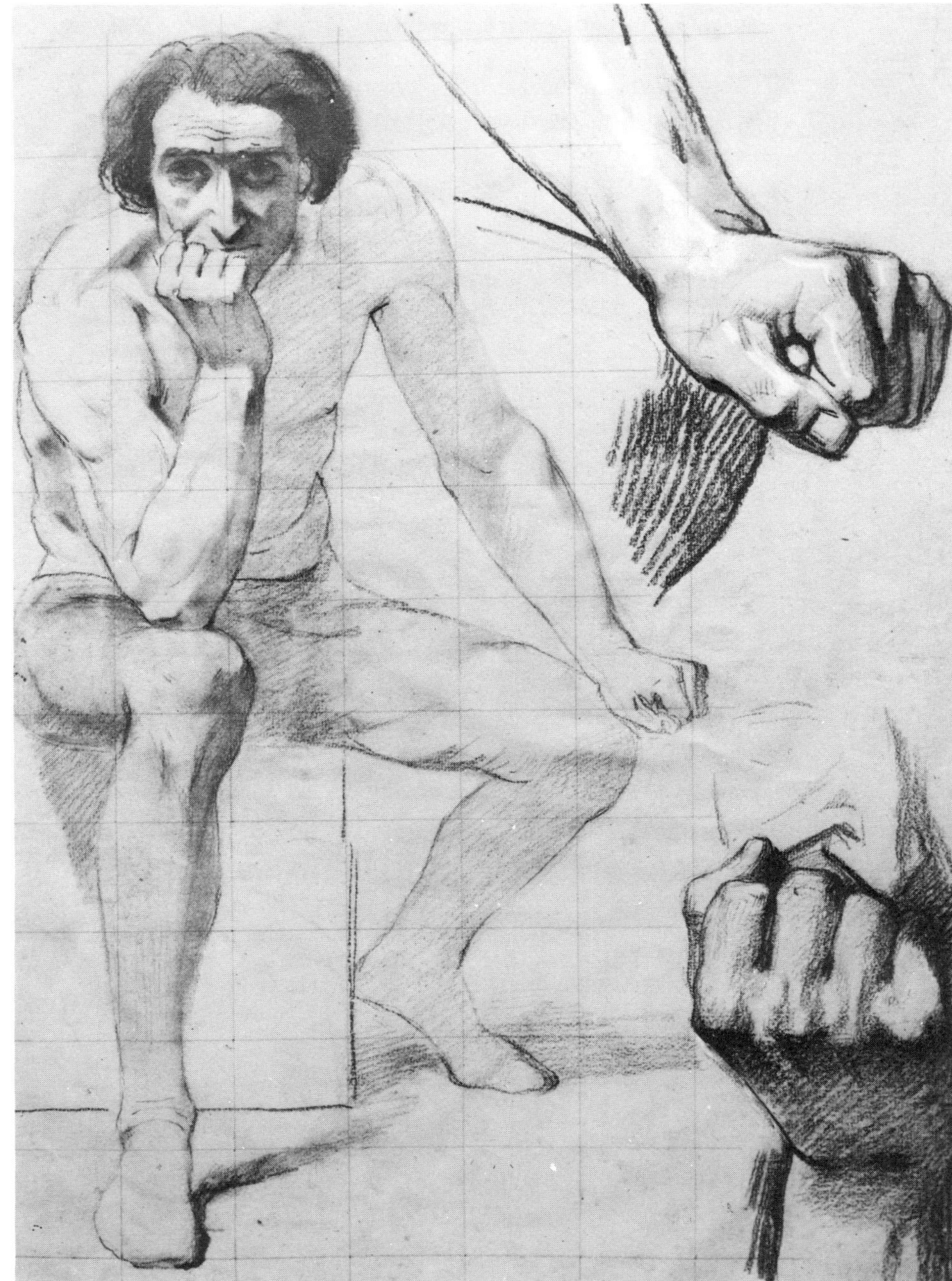

Black Crayon,
heightened with White
MUSEE DE LILLE, LILLE, FRANCE

HANS HOLBEIN THE YOUNGER (1497-1543)

Holbein the Younger was born in Augsburg, Germany. His father, Hans Holbein the Elder, was a much respected artist and it was under the tutelage of the father that this great German Renaissance artist received his schooling. Holbein's early works were devoted to church paintings which, unfortunately, have been destroyed. We know of their existence only through the drawings that he made as preliminary studies for the finished paintings. It was the Reformation that changed this master's dedication from church painting to portraiture. In the latter area he was one of the greatest masters. In 1533 he was appointed court painter to Henry VIII of England. The king provided him with a studio in the tower of Whitehall Palace, and Holbein was given a salary of £20 a year for his services. His talents were fully exploited in the field of painting royal portraits and miniatures, the designing of royal jewelry and interior decorations for the palace of the king. At the insistence of King Henry VIII, Hans Holbein was even made part of the delegations that were sent abroad to negotiate marriages for the king, his talents being employed to paint portraits of those who vied for the position of prospective brides. Finding himself literally overwhelmed with commissions for portraits, Holbein no longer could take time to work from the model. Instead, he was compelled to use a group of concise sketches which he used later to form the basis for his finished portraits. It was during the eleven years from 1532-1543 that he produced 150 portraits (in addition to many other works). His marvelous collection of portraits of English court figures are to this day unmatched. The plague, which was raging in London at the time, caused his death in 1543.

These three studies of hands are most likely done of the artist's own hands. The upper one shows his basic approach to establishing the outline of a mass. Note the delicate value change in the lower hand and the exquisite modeling of the texture of the flesh.

Pencil Drawing
THE LOUVRE, PARIS, FRANCE

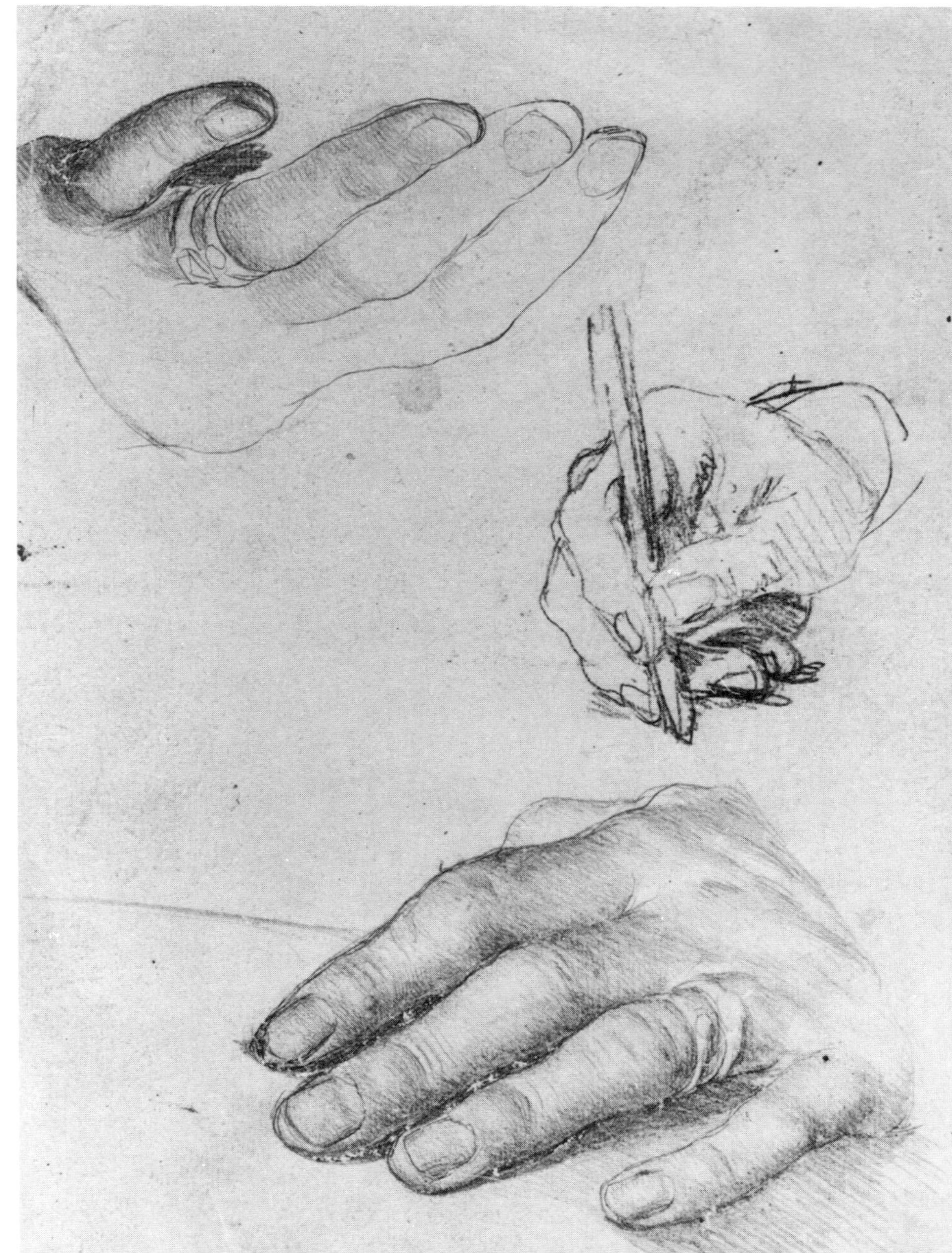

ALEXIS JOSEPH MAZEROLLE (1826-1889)

Alexis Mazerolle was a French historical muralist and portrait painter, student of Dupuis and Gleyre. The mural on the ceiling of France's National Theatre — Theatre Francaise — was done by Mazerolle. His work brought him the honor of being elected to the position of Officer of the Legion of Honor. His best known painting is *Nero Experimenting with Poison on a Slave* in the Lille Museum.

These black crayon, heightened with white, drawings were studies made for one of his mural commissions. The strength of the values (1) show that these studies were intended as models from which to paint. Details have been ignored for value pattern (2), however the details of structure where light revealed them have been clearly stated (3).

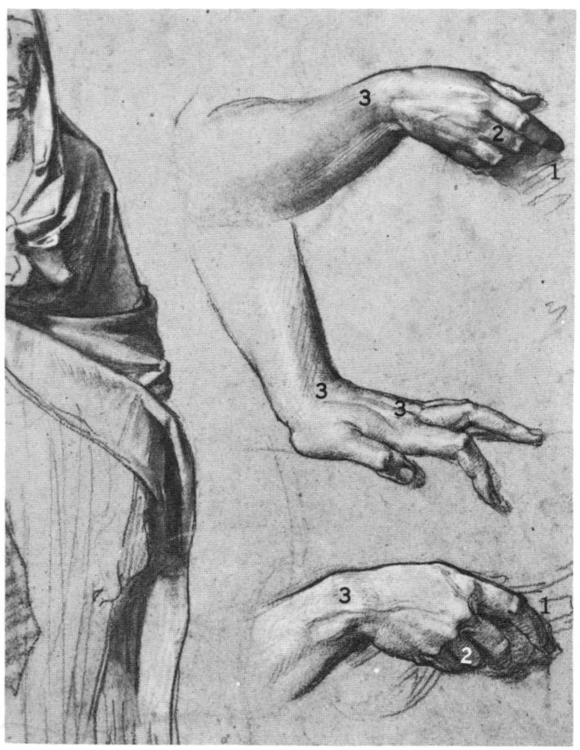

Black Crayon, heightened with White
MUSEE DE LILLE, LILLE, FRANCE

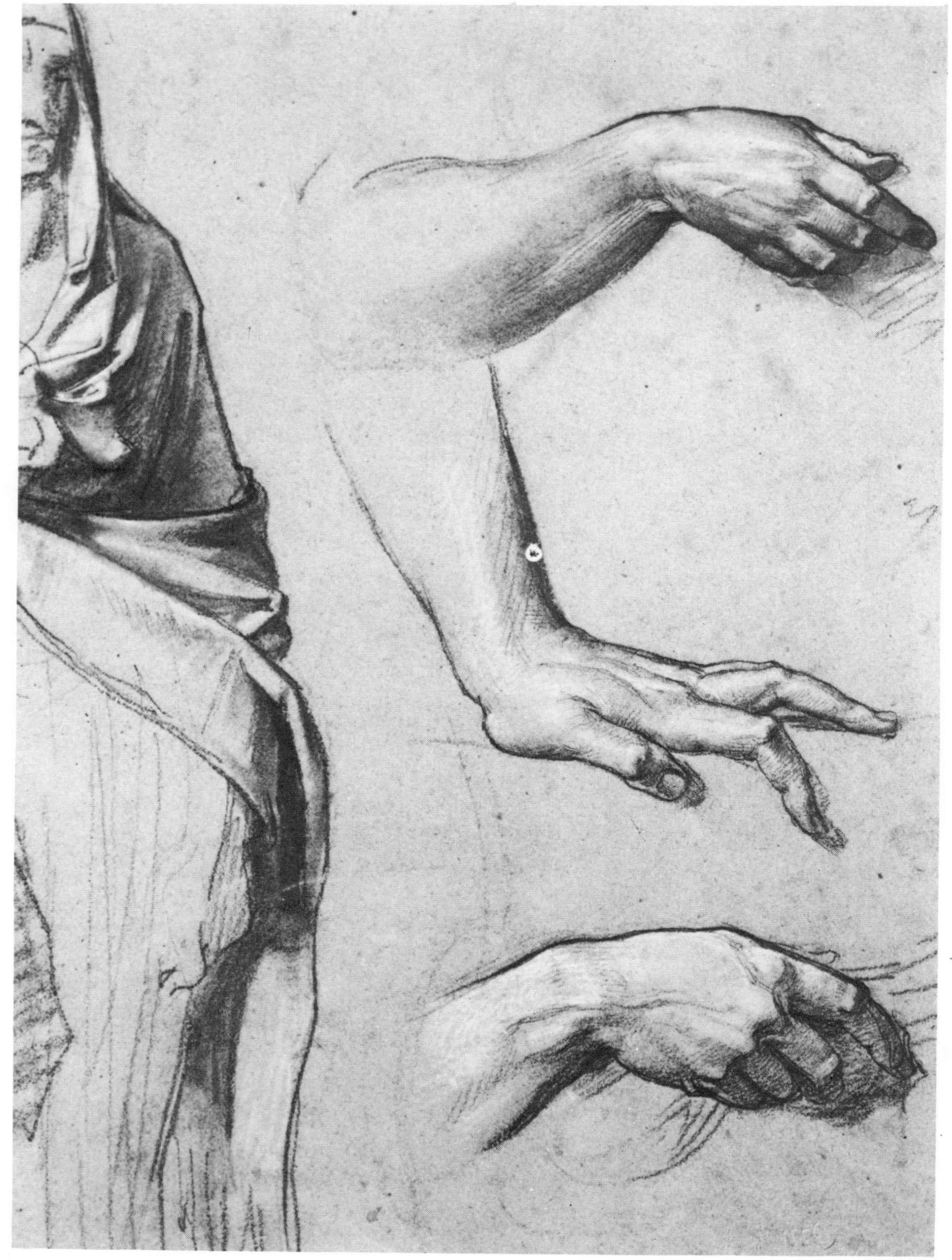

HENRY LEE McFEE (1886-1953)

McFee was born in Saint Louis. His early schooling in painting was in Pittsburgh, Pennsylvania. He later studied with Birge Harrison. McFee became interested in the modern movement of the cubists around 1911. He was also strongly attracted to the works of Cezanne. His first visit to Europe was made in 1921. Upon his return to the United States his work took on a more naturalistic quality in the mid-1920's. In the late 1920's and early 1930's, McFee's color became less brilliant and he developed a strong liking for muted hues. He painted, using sable brushes, and evolved a technique that was distinctly his own. The subject matter he preferred was mostly still life, however, he did many excellent figure paintings. He was elected National Academician in 1950. His awards were numerous. He was the founder of the Woodstock Art Colony. From 1942 he taught at Scripps College in Claremont, California. He died in Pomona, California.

In this pencil drawing one can see clearly McFee's mannered technique. An involved cross-hatch is used in the shadow pattern on the hands and also in background values. Line is of secondary importance in this drawing.

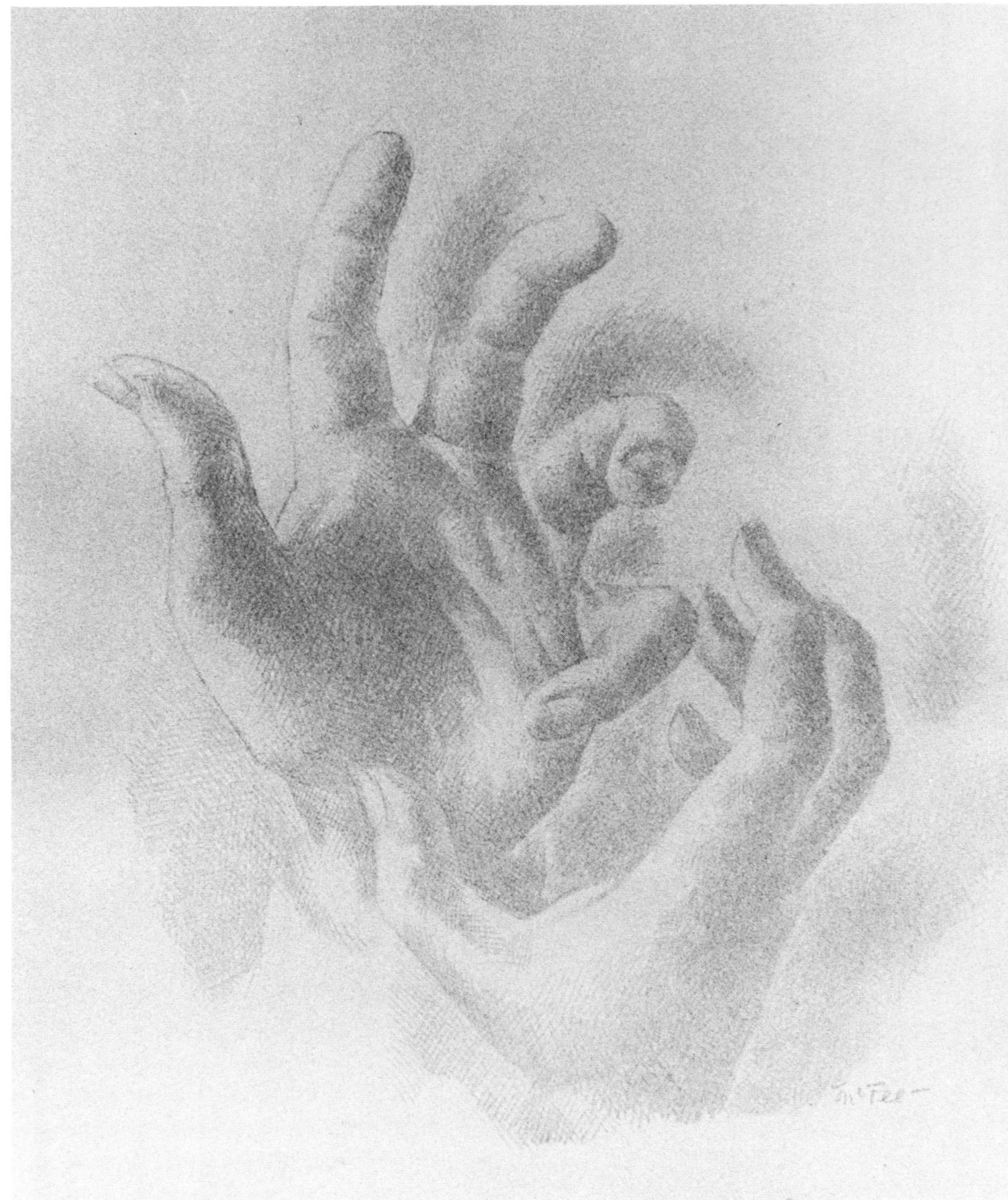

Pencil

COURTESY OF THE BOWDOIN COLLEGE MUSEUM OF FINE ARTS, BRUNSWICK, MAINE

GEORGE WESLEY BELLOWS (1882-1925)

Bellows was born in Columbus, Ohio, and educated at the Ohio State University. His interest and ability as a baseball player nearly resulted in his becoming a professional player in that sport. In 1904 he moved to New York where he became a student of Robert Henri. In 1910, 1918 and 1919 he was an instructor at the Art Students' League in New York City. He also taught at the Chicago Art Institute during 1919. His work was distinguished by strong composition, low keyed color and vitality of brush work. In his later work his color became more vibrant. In both his painting and his lithographs one finds works devoted to sports events from boxing to polo. His "Stag at Sharkey's" and his "Dempsey and Firpo" are two of his most famous. Portraits of his family, his mother, his aunts, his wife, and his children are prized possessions of American museums. His lithographs of revival meetings, jails, old-folks asylums, boxing matches, landscapes, urban subjects and members of his family show his interest in varied subject matter. He died at the age of forty three from a ruptured appendix.

This drawing made with conté crayon, was a study for Bellows' famous "Dempsey-Firpo Fight" which he did both in lithograph and in oil. The direct clean line of the lithographer's technique is clearly demonstrated here, also the strong pattern of the light values. Bellows was a ring-side spectator at this championship fight and drew from direct knowledge of the event.

Conté Crayon
PRIVATE COLLECTION

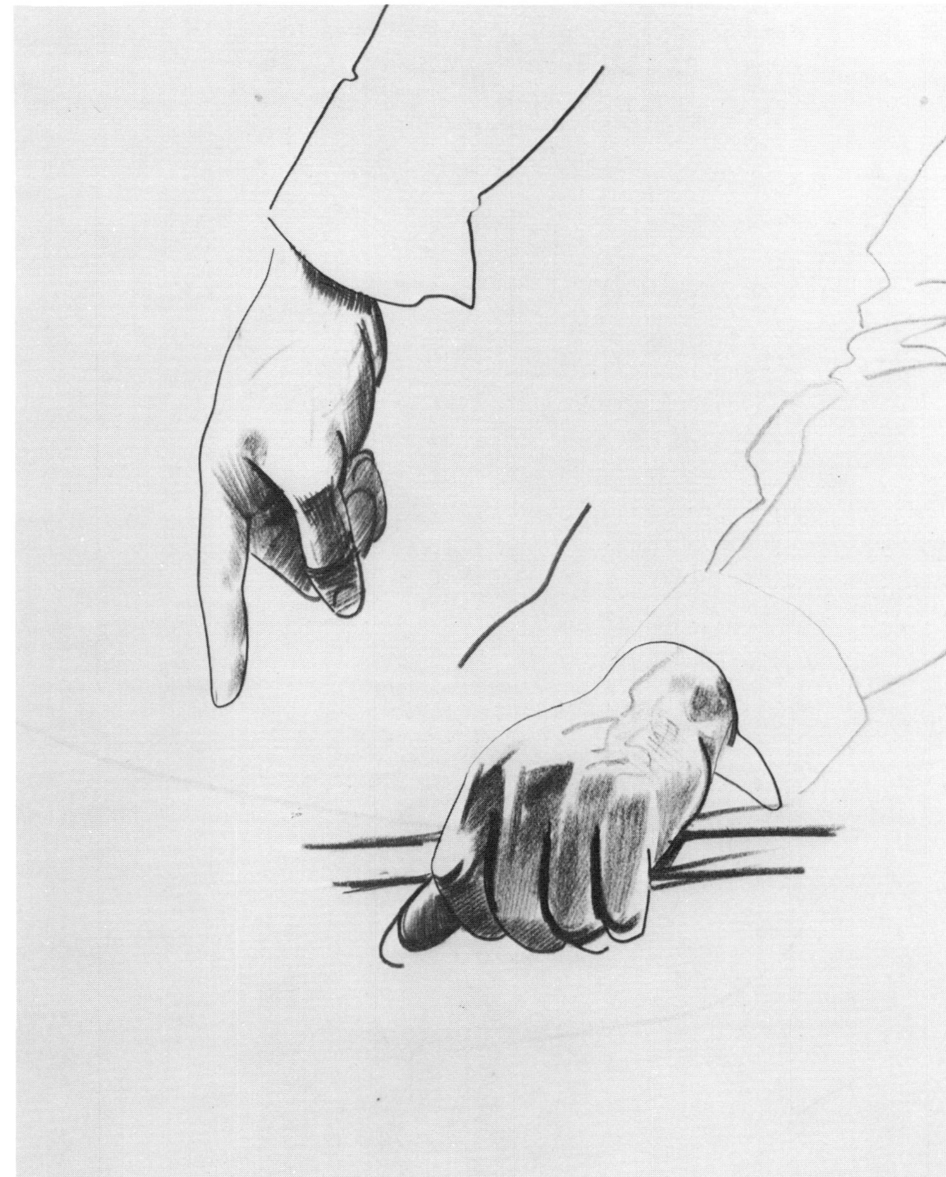

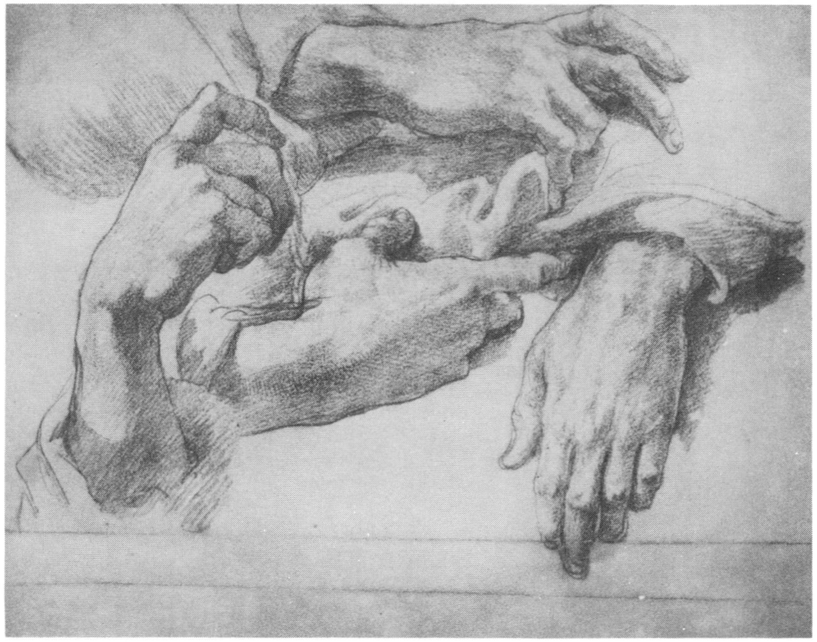

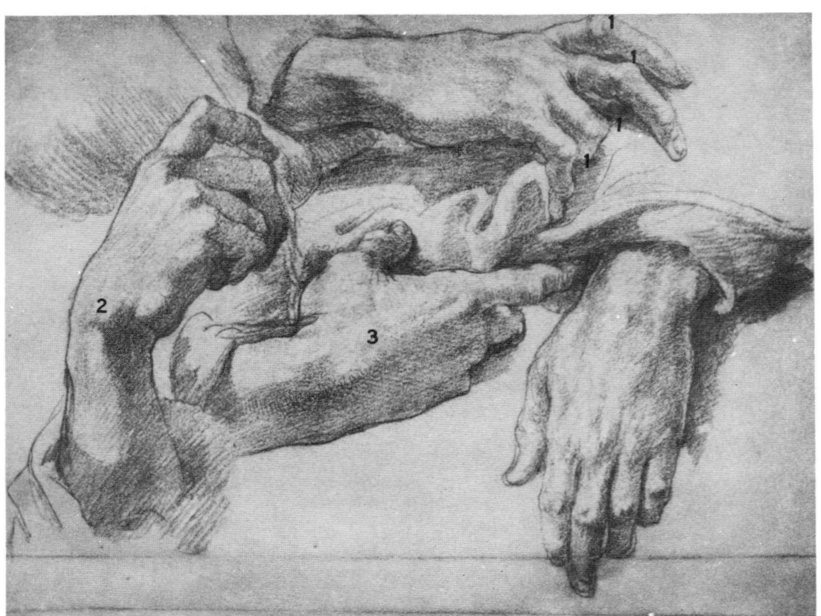

ANDREA DEL SARTO (1486-1531)

Andrea del Sarto was born in Florence in 1486, and there he had his first lessons in the rudiments of art from a modest painter, Andrea Barile. He then studied under Piero di Cosimo, and established a partnership with his fellow pupil, Franciabigio. He was little influenced by his master's art, which still belonged to the 15th century. Andrea preferred to study Leonardo, whom he was fully capable of appreciating, and he adopted the subtle shading of Leonardo's sfumato technique. He was also mindful of Raphael's harmonious composition, and of the strength of Michelangelo's forms. In his works the whole scene is always bathed in a soft light, and the sinuous line of an intellectual mannerism can be discerned. This is chiefly apparent in his exquisite drawings. From 1508 Andrea was a member of the Guild of Doctors and Apothecaries, to which painters were attached. In 1517 he married a beautiful widow, Lucrezia del Fede, who was often his model. Between 1518 and 1519 Andrea went to France at the invitation of Francis I. In 1530 during the siege of Florence he was commissioned to paint the portraits of the mutinous captains and traitors hanged in effigy. Andrea del Sarto died prematurely on January 22, 1531, at a time when Florence was in the grip of panic because of the plague. His death may have been due to this disease.

In these drawings by Del Sarto there is an interesting observation on the part of the artist at (1) — apparently the model suffered from arthritis — note the abnormal size of the knuckles. Hands (2) and (3) show the evidence of influence of both Michelangelo and Raphael.

Sanguine Crayon

UFFIZI GALLERY, FLORENCE, ITALY

VINCENT VAN GOGH (1853-1890)

It can safely be said that no painter has had more written about his mental health by psychiatrists than this mentally tormented Post-Impressionist. His early works were painted in low keyed browns and greens, the most famous of which was his *The Potato Eaters* (1885). His schooling in painting was short and undistinguished. In 1886 he went to Paris to join his brother, Theo, who was employed by the art dealer Goupil. Through his brother he met some of the Impressionist and Neo-Impressionist painters. He was induced to abandon his earthy colors for the pure and brilliant palette of divisionist technique of Seurat. He studied Japanese prints and the works of Delacroix and Monticelli. After two years in Paris he moved to Arles in southern France. Unable to sell his works, he was compelled to rely upon his brother for financial aid. In the bright sunlight of Arles he painted canvasses in pure brilliant color, thickly applied and with nervous strokes. Though he was committed to asylums periodically, he continued to paint. He invited Gauguin to join him at Arles where the two painted together a short time. In a fit of anger he threatened to kill Gauguin with a butcher's knife. He was again committed to the care of Dr. Gachet and released after a few months. While painting one day he took his life with a revolver shot through his stomach.

Van Gogh has never been thought of as a draughtsman. In these drawings of a sower's hand Van Gogh's line shows the nervous energy and directness of his technique in painting. The character of the peasant's hand is strongly stated but not beautifully drawn. There is more than a hint of Oriental influence in Van Gogh's line.

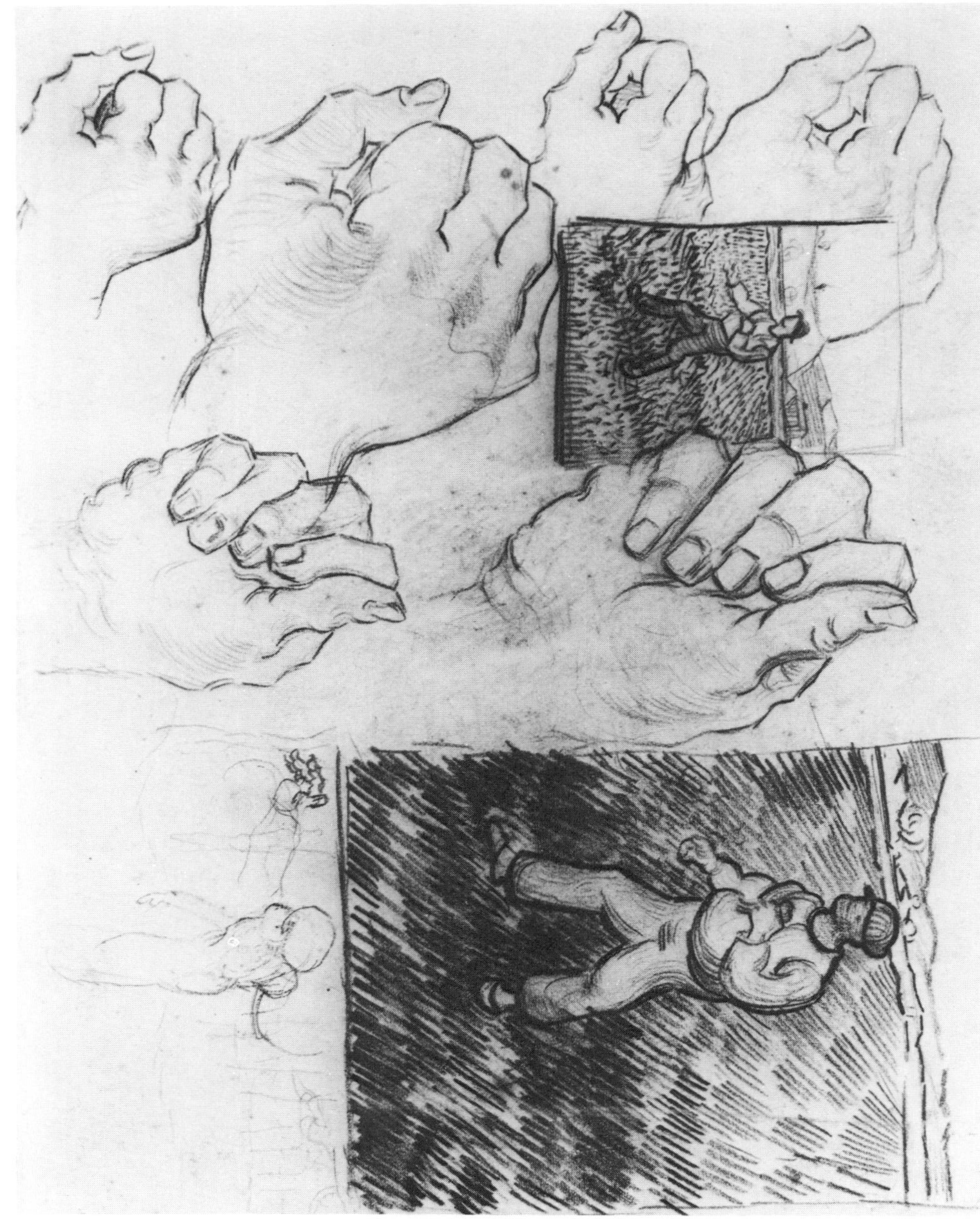

Black Crayon
THE MUNICIPAL MUSEUM OF AMSTERDAM, HOLLAND

JEAN AUGUSTE DOMINIQUE INGRES (1780-1867)

French classical painter. Ingres' father was a sculptor of little distinction. Ingres' first interest was music — he was sufficiently accomplished as a violinist to play Viotti's concerto to win acclaim as a boy. His father enrolled him in the Academy of Fine Arts in Toulouse where he saw a copy of Raphael's *Madonna della Sadia* that convinced him that he wished to become a painter. At the age of sixteen he entered the studio of David, France's most honored artist of that time. In 1801 he won the Prix de Rome; however, the affairs of France were so confused at that time that he was not able to go to Rome until 1806. His years in Rome were not easy and he earned his living doing pencil portraits which to this day are considered small masterpieces. He revered the paintings and drawings of Raphael and attempted to emulate the great Italian. His success upon returning from Rome was sporadic. In 1825 he was elected to membership in the Insitute and from that date his success was assured. His greatest work was in the field of portraiture. His classical historical subjects were cold, lifeless but technically perfect. He was worshiped by Degas, Paris de Chowin and Gauguin. He was also greatly admired by the Cubists, strange as that may seem.

In this drawing you will note that Ingres made four attempts at the contour of the hand at (1). The character of these hands is particularly French of that period, dimpled at the knuckles (2), full and round of form in the fingers. Ingres was a great admirer of the Renaissance draughtsmen, particularly Raphael. Their influence is found in the type of line Ingres used. Degas was an admirer of Ingres and Ingres' influence is found in Degas' early drawings.

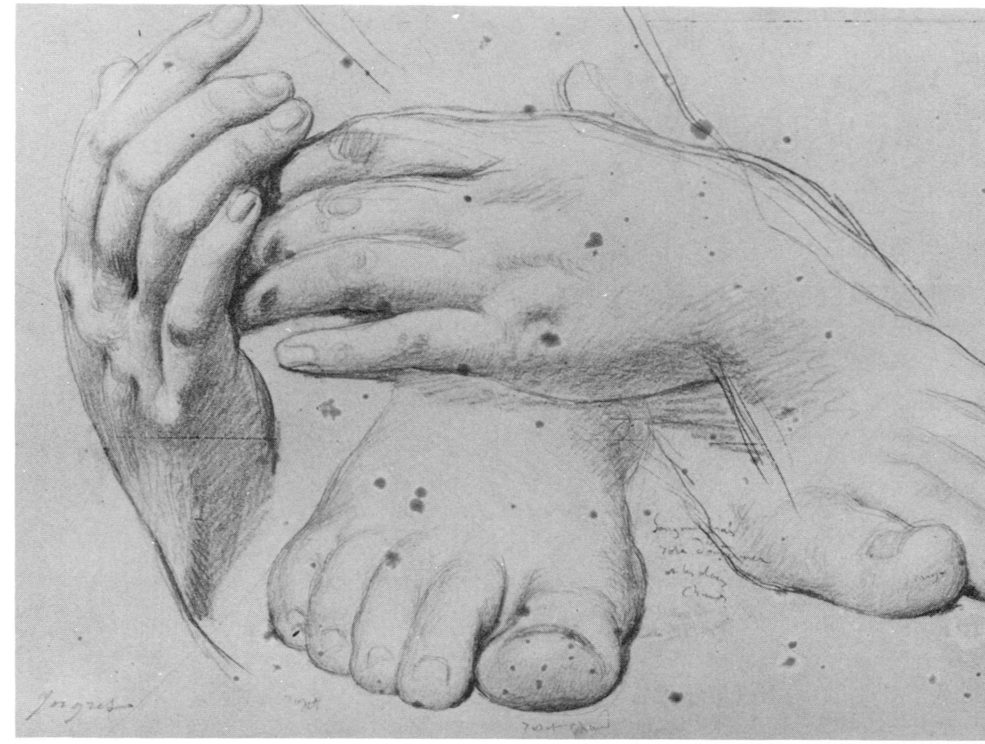

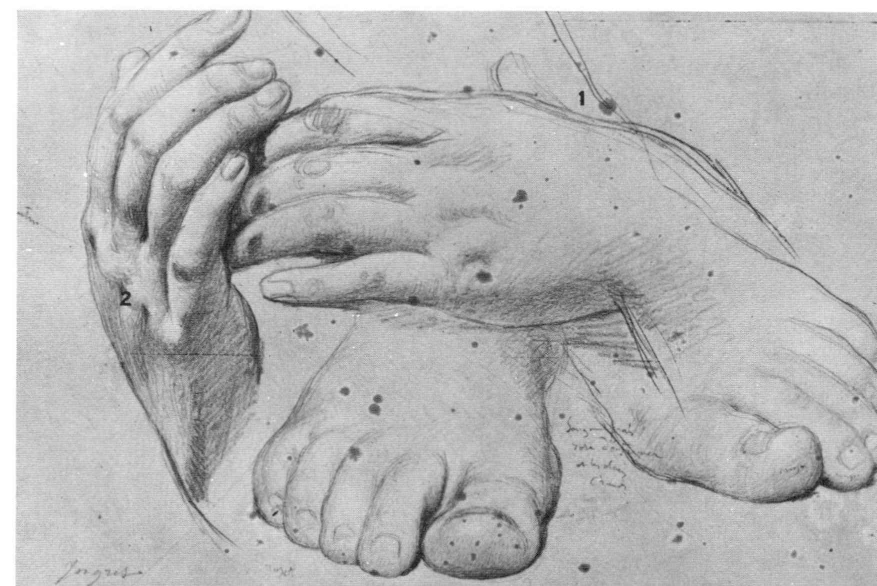

Pencil

MUSEE INGRES, MONTAUBAN, FRANCE

BENJAMIN ROBERT HAYDON (1786-1846)

Benjamin Robert Haydon is known as a distinguished English painter of historical subjects. His father was a prosperous printer and publisher, who, sensing the inherent gift in his son, Benjamin, enrolled him in the Royal Academy in London to pursue his studies in art. Here he exhibited the great talent that his father had foreseen. However, Haydon's bad temper coupled with his quarrelsome inability to get along with his fellow artists brought him ill luck which seemed to dog his footsteps throughout his life. When he was forty years of age he was committed to prison for debts — he was again imprisoned in 1827 for the same reason. While in prison he painted the *Mock Election* which was bought by King George IV for the comparatively large sum of £500. While working on his last painting, *Alfred and Trial by Jury,* he was further pestered by continuing debt and disappointment. Despondent because of all the harsh treatment life had dealt him, he committed suicide by pistol-shot, leaving a note in which he said, "stretch me no longer on this rough world".

In examining this drawing it appears that Haydon forgot to draw the indication of the index finger above the thumb. Further study shows that the model's arm was above the eye-level of the artist and that the thumb hid the index finger. The modeling of the fore-arm shows that the arm was above the artist's eye-level by the direction of the lines at (2).

Black Chalk, heightened with White
THE BRITISH MUSEUM, LONDON, ENGLAND

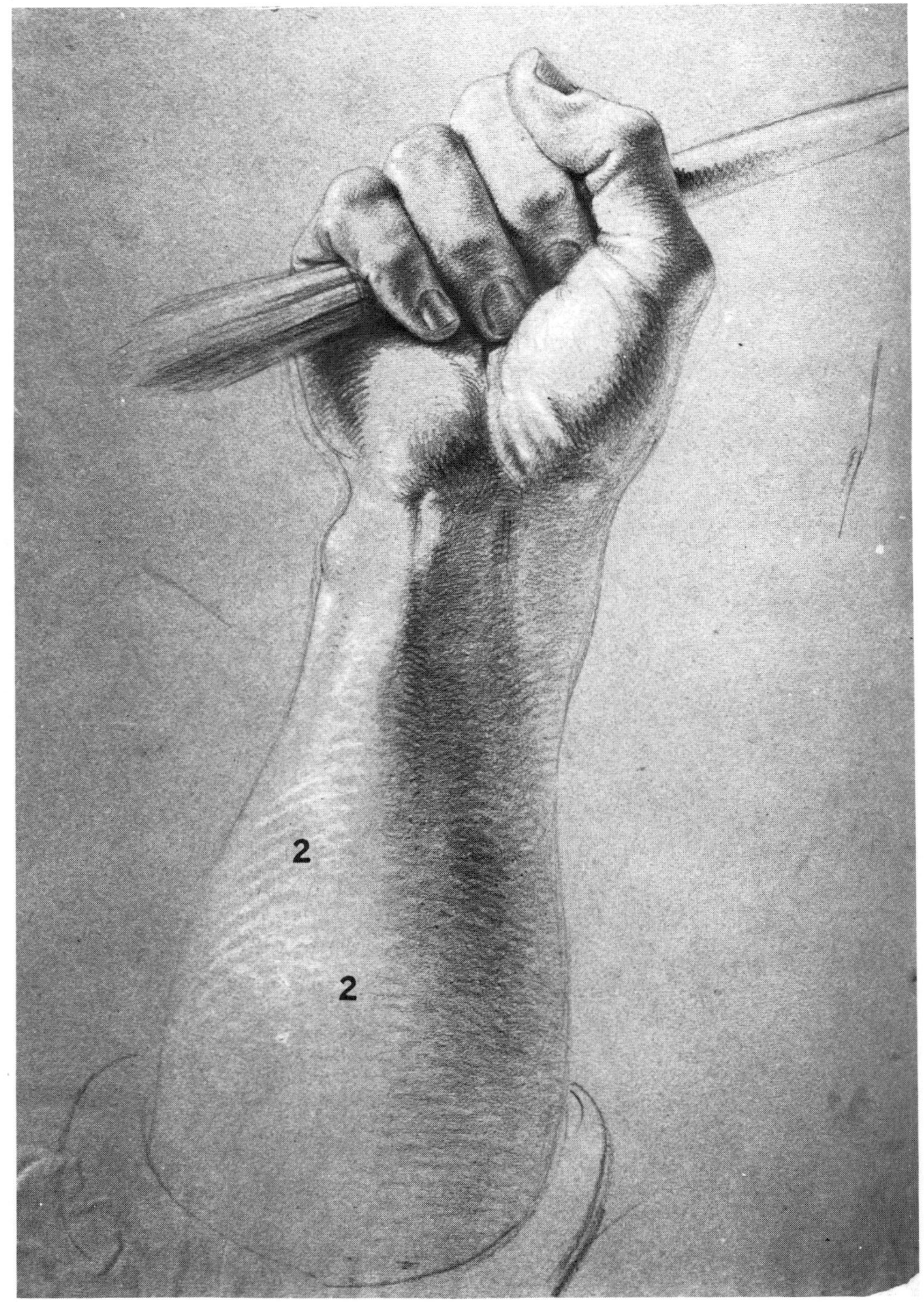

LORSER FEITELSON (1898-)

Feitelson is a contemporary American painter. His interest in art was awakened in his father's library which was replete with books and magazines on art. He studied with the sculptor Arthur Tefft and George Bridgman. Feitelson is best known among critics as one of the early leaders of the "Hard-edge Painting School", having explored this mode of painting as early as the mid-1930's. Works of his in this school of painting are to be found in museums and private collections. Curiously, Feitelson is a profound student and noted collector of the drawings of the Renaissance draughtsmen. His collection of drawings, engravings and etchings of the great master draughtsmen is enviable. His wife, Helen Lundeberg, is a contemporary painter with paintings of note in museums and private collections. Feitelson is Professor Emeritus and former Chairman of the Department of Fine Arts at Art Center College of Design. He is a distinguished draughtsman and an ardent teacher of drawing in the traditional manner.

This is a reproduction of a demonstration drawing made by Feitelson before his students at Art Center College of Design. Drawn on toned charcoal paper, Feitelson has allowed the tone of the paper to represent the intermediate value and drawn the highlights with white chalk. Both drawings are intended to show the rhythm of the gesture of the hand and to indicate the salient anatomical structure.

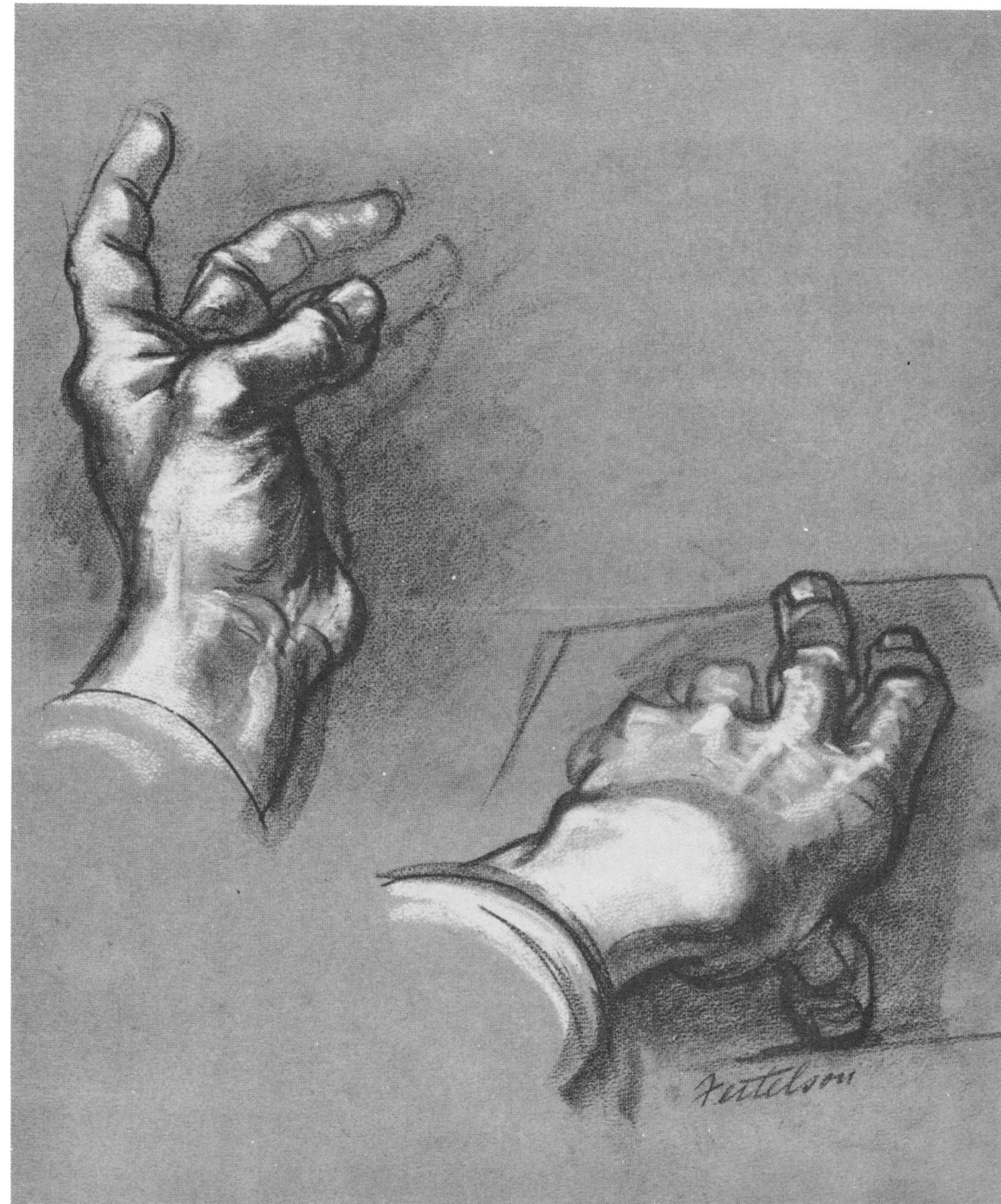

Charcoal, heightened with White Chalk
LORSER FEITELSON COLLECTION, LOS ANGELES, CALIFORNIA

PABLO RUIZ PICASSO (1881-1973)

Picasso's father was José Ruiz Blasco, a well-known teacher of drawing and painting at the Academy in Barcelona and his mother was a Castilian — Maria Picasso. By the age of sixteen Picasso was so accomplished a draughtsman that he left the Academy in Barcelona. His early paintings were so influenced by Toulouse-Lautrec that there was difficulty in determining whether they were not works of the French painter. Picasso was a great innovator of technical approaches to painting. He and Bracque were the great developers of Cubism. Picasso restlessly moved from one technical approach to another. He has had greater influence upon young painters than possibly any other painter. He is without question the best known and the most controversial painter of this century.

This drawing is from Picasso's "Blue Period" — when all of his paintings were executed in various shades of blue. The drawing is fluid in line and exaggerated in proportion. Though Picasso could draw with great exactitude, this was something that did not intrigue him. He was interested only in the establishment of the basic gesture during these years. The drawings have been made in a calligraphic (written) style with no concern for light pattern or anatomical features.

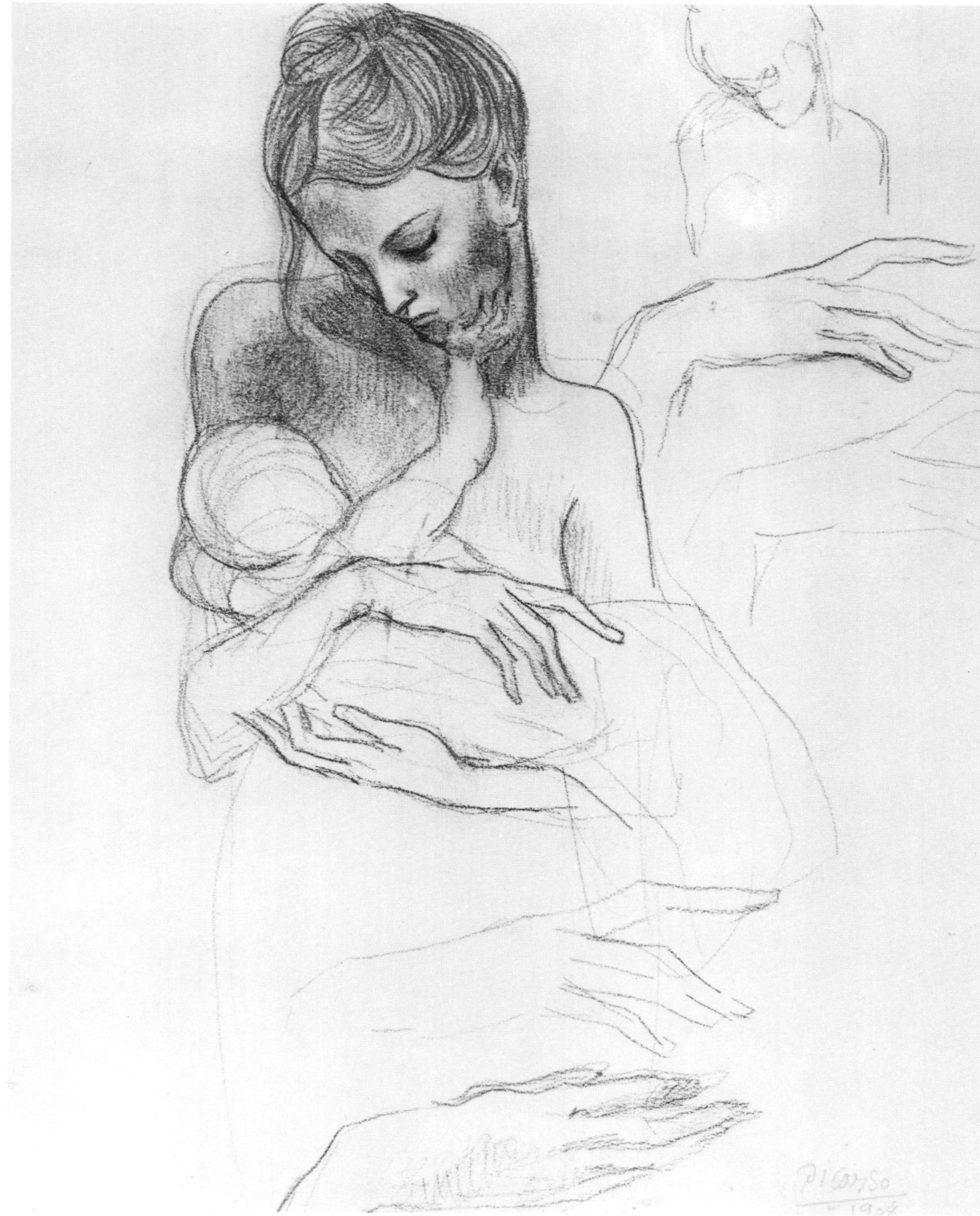

Black Crayon on Cream-Colored Paper
THE FOGG ART MUSEUM, HARVARD UNIVERSITY, CAMBRIDGE, MASS.

SIR EDWARD JOHN POYNTER (1836-1919)

Poynter was born in Paris. He attended the Royal Academy School in London and later studied with Gleyre in Paris. Poynter was a painter of historical subjects set in the ancient world. His technique was classical and his skill of a high order. His subject matter was in keeping with the demands of Victorian taste and he enjoyed great success in England. He was elected president of the Royal Academy and was the first professor of the Slade School of Art at University College in London. Poynter was a highly competent academic draughtsman.

This drawing was preparatory for one of his classical paintings and was squared off for enlargement (1). The various poses of the hands were studied for their slight changes of gesture (2) and (3). The anatomy has been strongly indicated (4). The light pattern has been studied but the major effort has been toward construction.

Pencil

VICTORIA & ALBERT MUSEUM, LONDON, ENGLAND

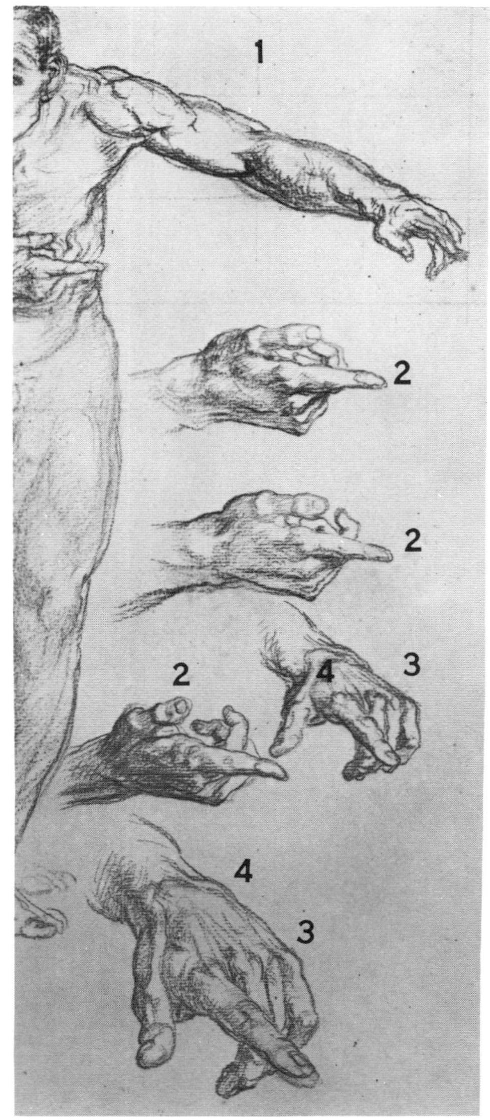

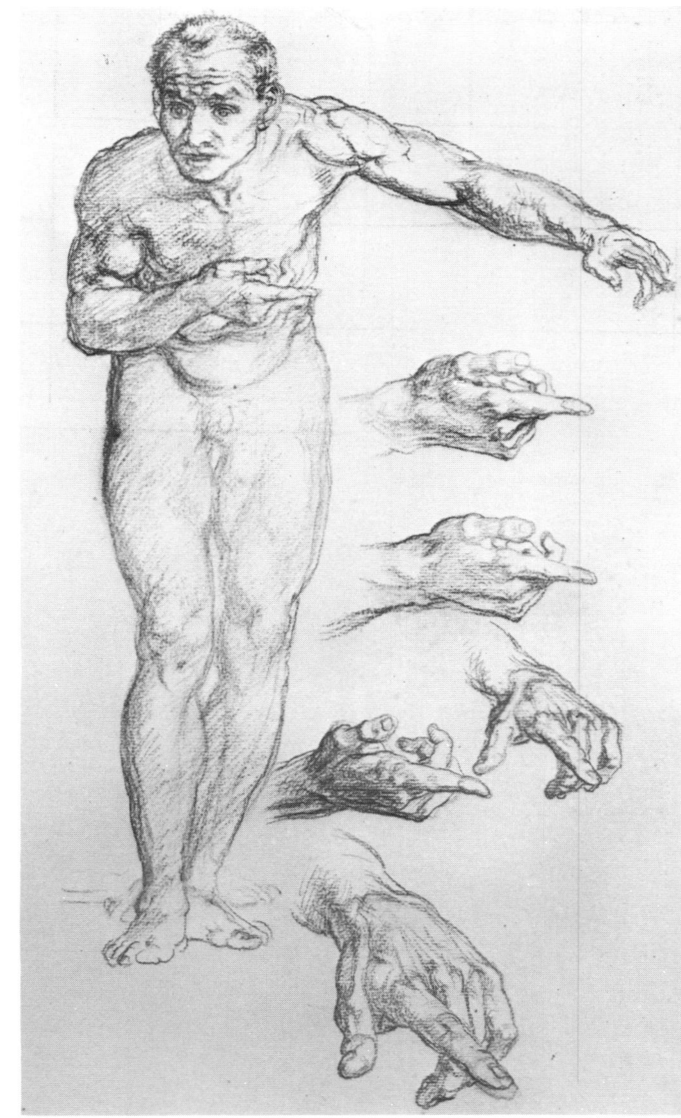

LEONARDO DA VINCI (1452-1519)

This extraordinary Renaissance man who was a painter, sculptor, architect, engineer, mechanician, musician and naturalist was the illegitimate son of a Florentine lawyer. He was unusually handsome and gifted with great personal charm. His intellectual curiosity and energy were inexhaustible. He entered the studio of Andrea del Verocchio who was his only teacher. Leonardo was a seeker for knowledge in every field and initiated study in areas previously unexplored. His studies in theoretical and applied science were, for his day, monumental. His architectural and engineering projects amazed his contemporaries. It is written, that even as a youth, he would seclude himself for long periods of time in intellectual absorption totally unmindful of rest or food. He was the friend of all of the great minds of his day. His company was sought as a skilled musician, as a reciter of poetry, as a horseman and for feats of strength. His notebooks, which were filled with drawings and written observations of flights of birds, rocks, mountains, water, cloud effects, etc., are still studied by scholars. His services were sought by popes, nobles and kings. His last days were passed in the service of Francis I of France. His best known paintings are the *Mona Lisa* and *The Last Supper* in Sta. Maria della Grazie in Milan, Italy. It is reported that he said during his last years, "Had I devoted myself to painting I think that I could have become one of the great painters." He died in Amboise, France, May 2, 1519.

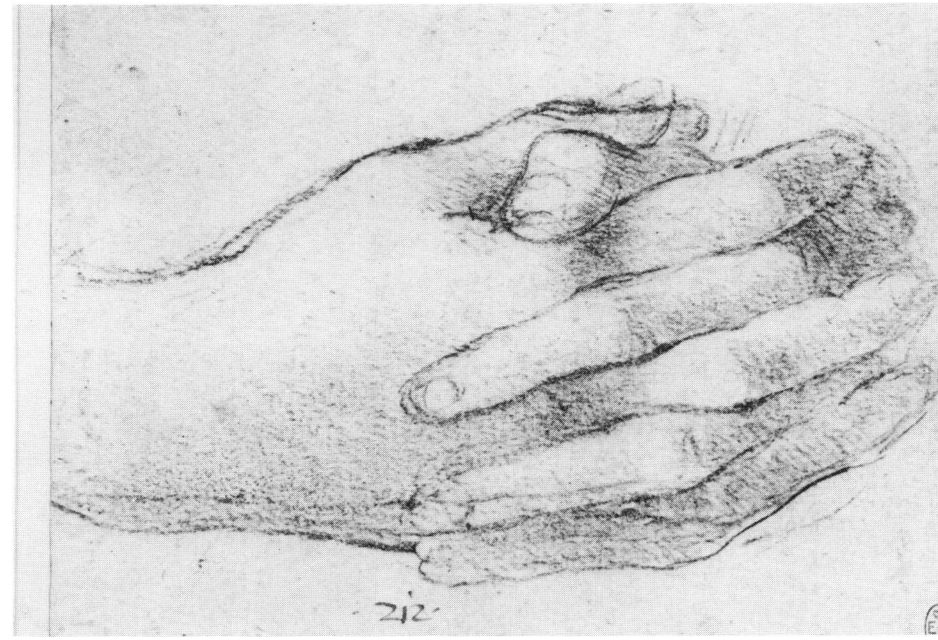

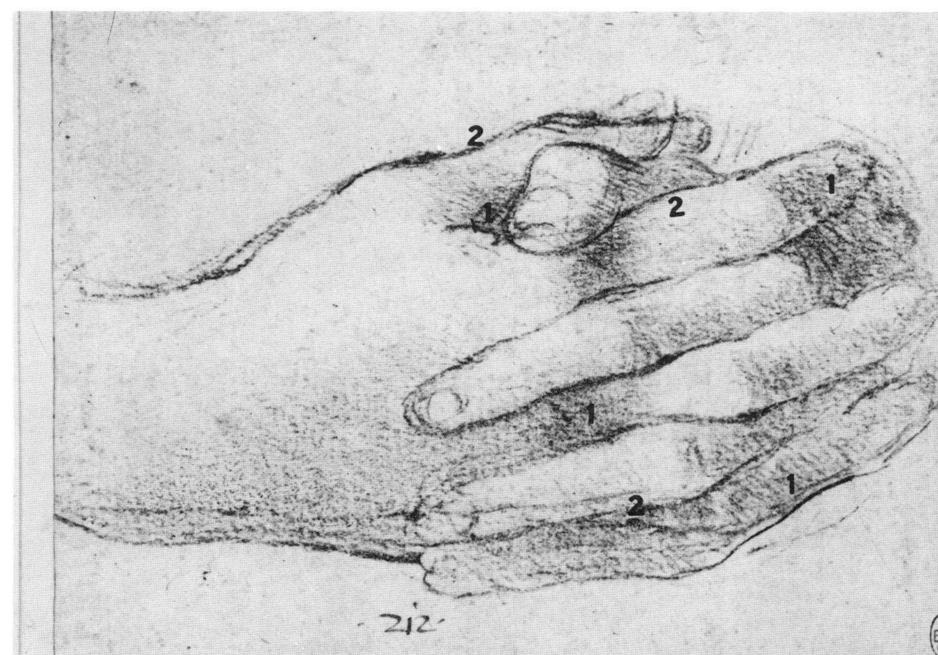

Leonardo was the first painter to recognize the play of light and shade as one of the most significant elements in painting. In this drawing he was as conscious of this quality of light and dark as he was of the line (1). The sensitivity of the line (2) has added to the attractiveness of the drawing as well as defined the contour of the fingers.

Black Chalk

ROYAL COLLECTION, WINDSOR CASTLE, BERKSHIRE, ENGLAND
COPYRIGHT RESERVED

BEN SHAHN (1898-1969)

Shahn was born in Lithuania and came to the United States with his family when he was eight years old. He was apprenticed to a lithographer at fifteen years of age and was a commercial lithographer until the age of thirty two. He started studying painting at the National Academy of Design in New York City and later at the Grande Chaumier in Paris. He traveled and studied in France, Italy, Spain and North Africa. Shahn concerned himself with social issues as is evidenced in his works of "Dreyfus", "Sacco and Vanzetti", etc. In 1933 he was the assistant of Diego Rivera while that artist was painting his murals in Radio City in New York. Shahn was a Neo-Romantic. Shahn's paintings show an intention toward realistic qualities but are executed in flat harsh color and incisive line.

These drawings made with a brush and ink cannot be truly classified as drawings of hands but as a drawing of a gesture of despair in which the hands are the paramount expressors of the emotion. The jagged line in both drawings add to the feeling of anguish. In the upper drawing the figure has been compressed into a column of tight emotion while the lower drawing gives the feeling of the pull of the earth upon the figure. Note that in neither drawing does the head contribute to the strong emotion since in the lower drawing it does not exist at all.

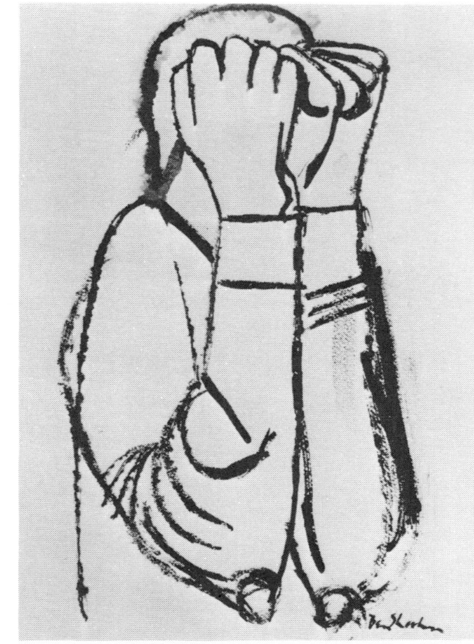

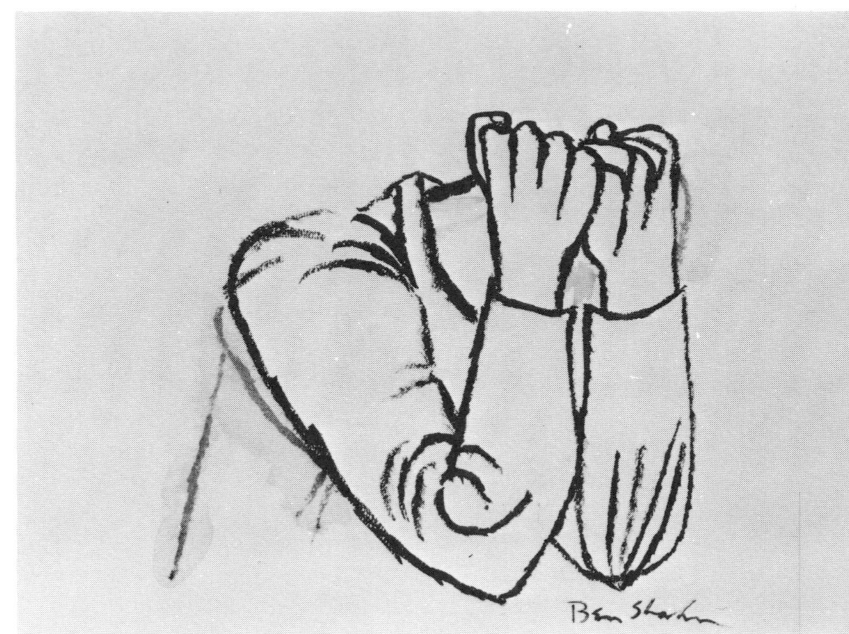

Pen and Ink and Wash
COURTESY OF THE DETROIT INSTITUTE OF ART, DETROIT, MICHIGAN

NICOLAS FRANÇOIS OCTAVE TASSAERT (1807-1874)

Tassaert's early life was unfortunate and difficult. Orphaned at a very young age, he was compelled to support himself when only twelve by working as an engraver and a lithographer. Admitted to the Royal School of Beaux Arts at the early age of seventeen, he soon showed unusual talent and became an exhibitor at the Salon at the exceptional age of twenty. He was a painter of historical subjects as well as expertly executed scenes from contemporary life, works which were extremely well researched for detailed accuracy, well drawn and carefully painted. He enjoyed considerable approval of the critics of his day and his works were always especially admired by the general public. Suddenly, his great genius was interrupted. At the age of forty five Tassaert's eyes began to fail and eight years later he was blind. For the remaining fourteen years of his life he vegetated, forgotten by critic and public alike. Despondent over the blow that fate had dealt him, he committed suicide by asphyxiation, and thus his genius was stilled forever.

This drawing, done in sanguine chalk, is a good example of a full scale of values to record the play of light on a pair of hands. Note the indication of the tendons of the extensors of the fingers (1). The grace of the gesture of the hand (2) has been well stated. Compare the quiet, soft quality of these hands with the rugged, tormented drawing of the hands on the opposite page by Ben Shahn.

Pencil heightened with White and Red Chalk
THE LOUVRE, PARIS, FRANCE

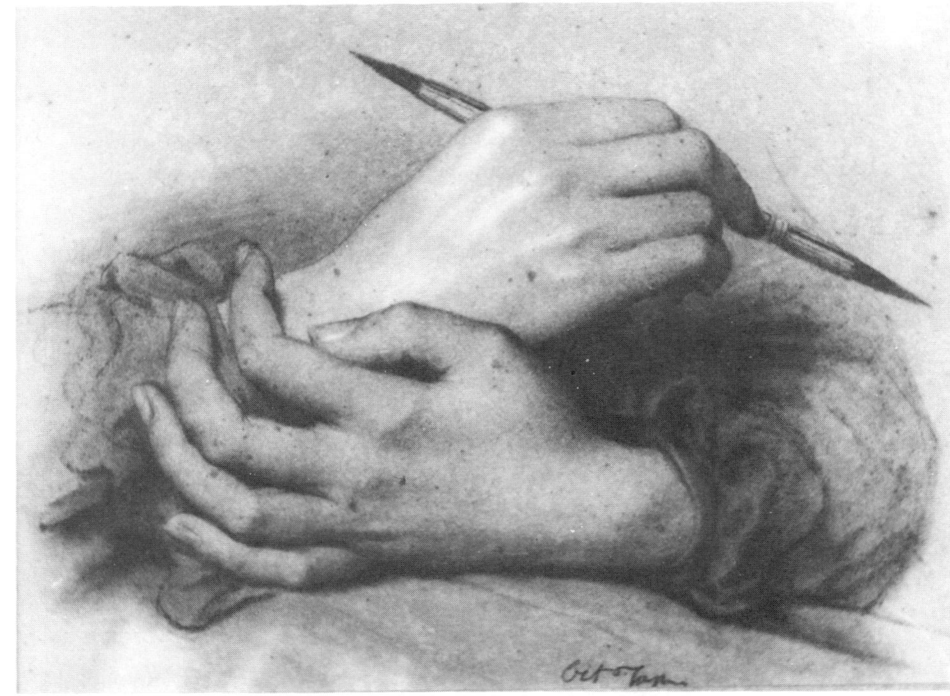

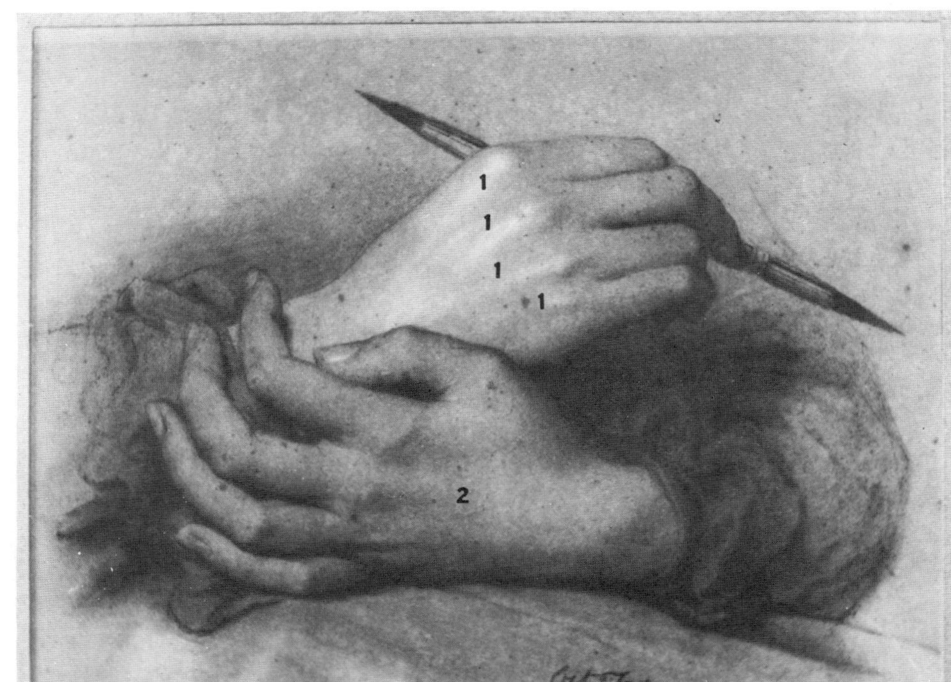

JOHN SINGER SARGENT (1856-1925)

Sargent was born in Florence, Italy. His father was a physician from Philadelphia who had gone with his wife to live abroad. Sargent had shown at an early age a leaning toward art which his mother encouraged. His early pencil drawings of landscape show a facility that belied his age. Sargent started his study with Carolus-Duran in Paris in 1874 at eighteen years of age. Carolus-Duran was a great admirer of the works of Velasquez and it was from this master that Sargent got his great love of the Spaniard's works. Carolus-Duran was a strong advocate of "alla prima" painting, allowing his student no preparatory drawing on the canvas prior to painting, but demanding that the planes be directly painted with large, wide brushes. Sargent, without question, was still his most brilliant student, winning honors early. A painting of *Mme Pierre Gantreau* (*Mme X* now in the Metropolitan Museum, New York) brought Sargent some unwanted publicity and he left Paris and established himself in London where he became a huge success. The great and the wealthy of Europe and the United States beat a path to his studio to have their portraits painted. Tired of painting "mugs", as Sargent called them, he abandoned portraiture for murals and water-color paintings in his last years. His water-colors are still looked upon as models of directness of execution and brilliance of technique. He lived in England for the remainder of his life making occasional trips to the United States for the purpose of executing commissions. He is generally considered the master portraitist of this century.

Sargent's favorite medium for drawing was charcoal as in this reproduction. A powerful man physically, Sargent drew with power — directly and with strength of line. Both his eye and hand were disciplined as can be seen in the unwavering directness of this drawing made as a study for his murals in the Boston Public Library.

Charcoal

CORCORAN GALLERY OF ART, WASHINGTON, D.C.

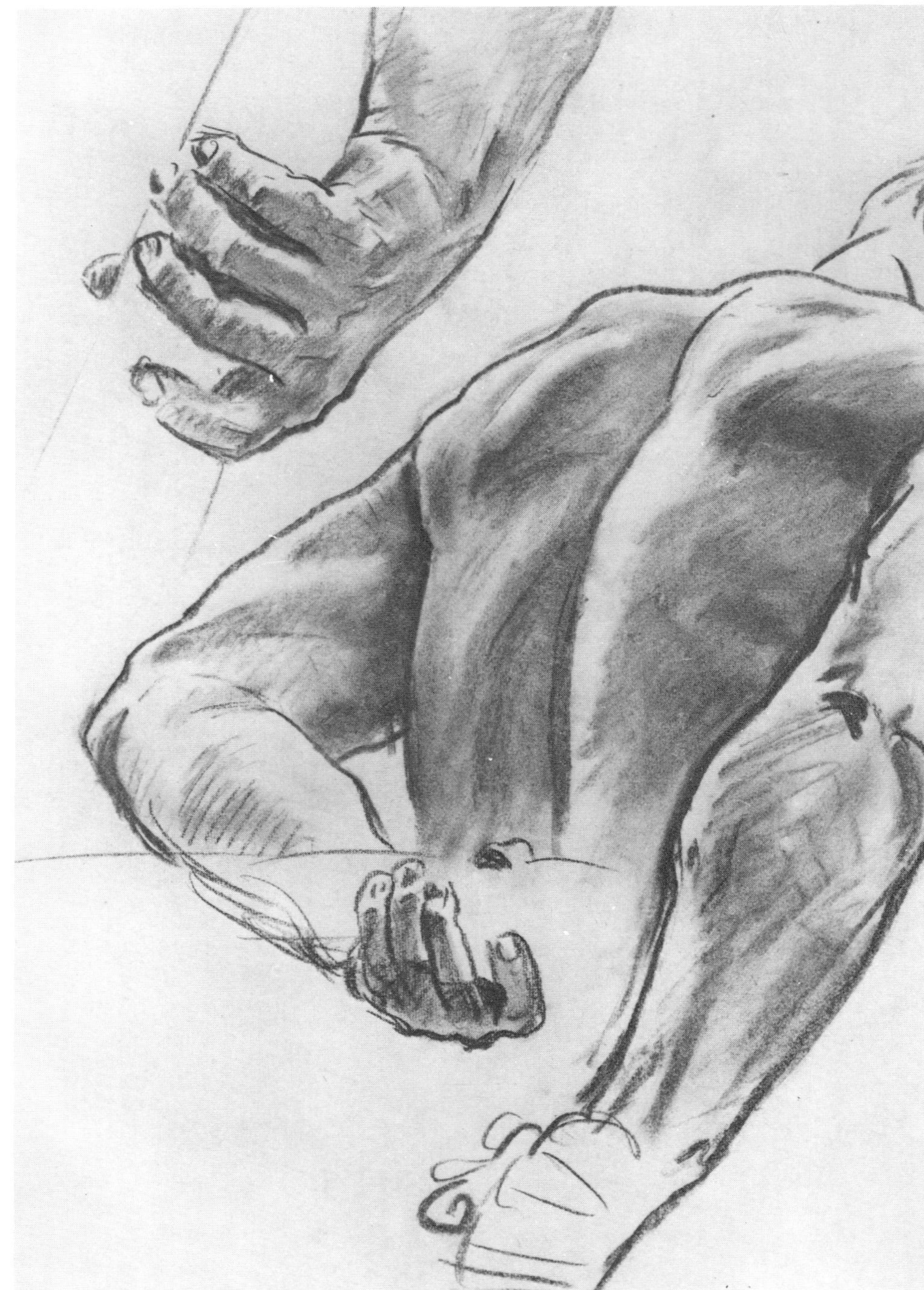

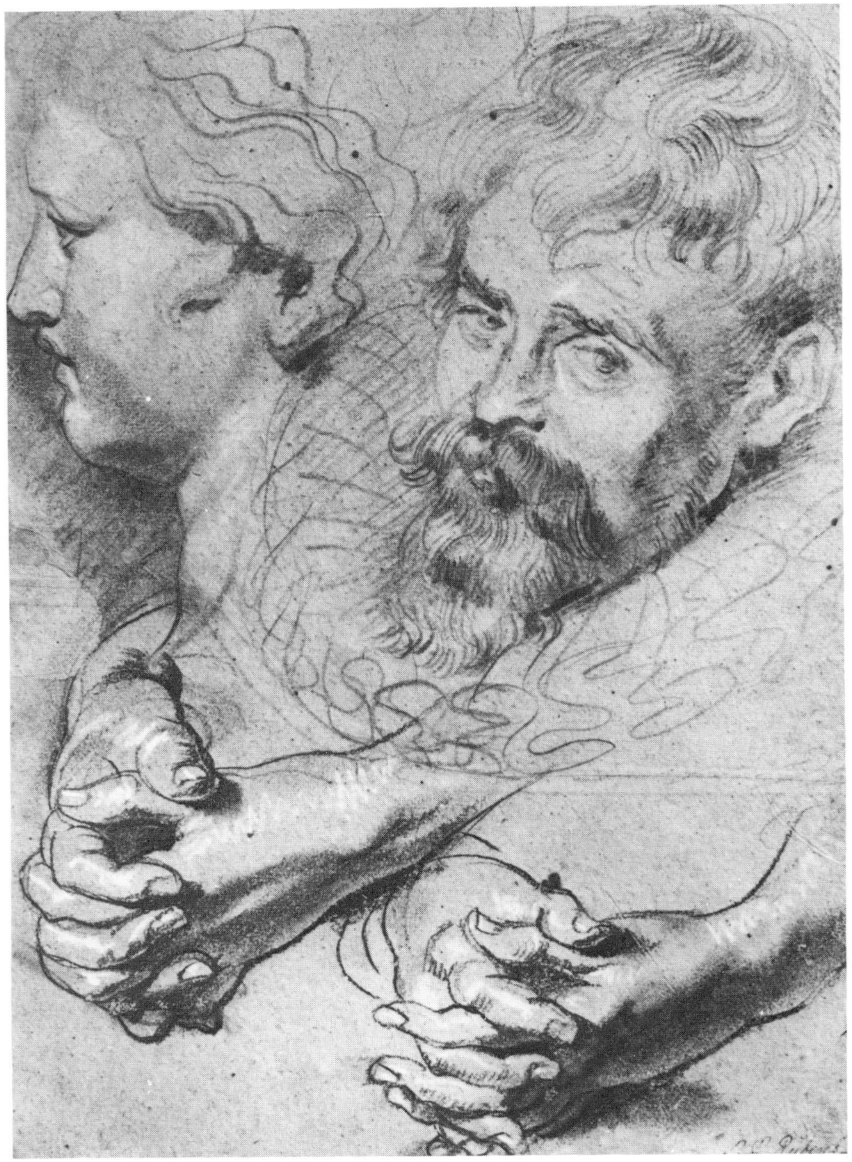

Black Chalk, heightened with White
ALBERTINA MUSEUM, VIENNA, AUSTRIA

PETER PAUL RUBENS (1577-1640)

This Flemish painter of Baroque style enjoyed a career unique among painters. The volume of work completed by him is almost unbelievable when one considers his other activities as Ambassador to the Kings of Spain and England. Rubens was granted honors and privileges generally reserved for royalty. He was an accomplished linguist, connoisseur of antiquities and a scholar of vast knowledge. When he died, he was mourned as "the most learned painter in the world." He was knighted by the King of England and was secretary of the King of Spain's Privy Council.

In these drawings of clasped hands note the strength of line and the directness of execution (1). There is no feeling of searching for the form — it is stated with the authority of knowledge — what the Italians called "alla prima", namely, put it down correctly from the first. The values of the darks have the strength of Rubens' own physical strength (2).

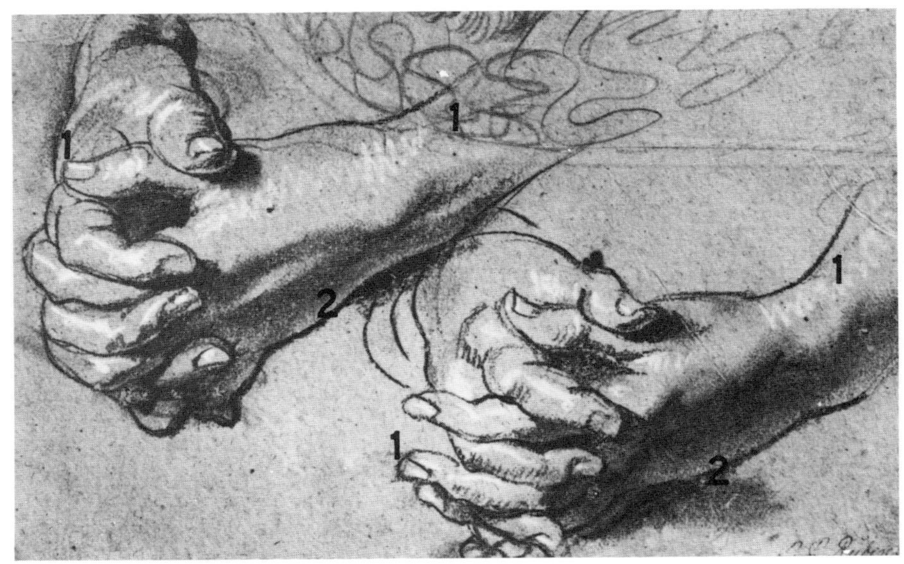

CORNELIS DE WAEL (1594-1662)

De Wael was born in Antwerp. His father, who was a painter, was his first teacher. Early in his life he and his brother went to Genoa, Italy. He received commissions to paint pictures for a number of churches in Genoa, however, the greater bulk of his work was devoted to painting military subjects, particularly battle scenes. The Duke of Arschot was one of his most ardent patrons as was Philip III of Spain who held de Wael in high esteem.

These drawings are excellent examples of economy of means necessary to establish the essentials for drawing the hand and arm.

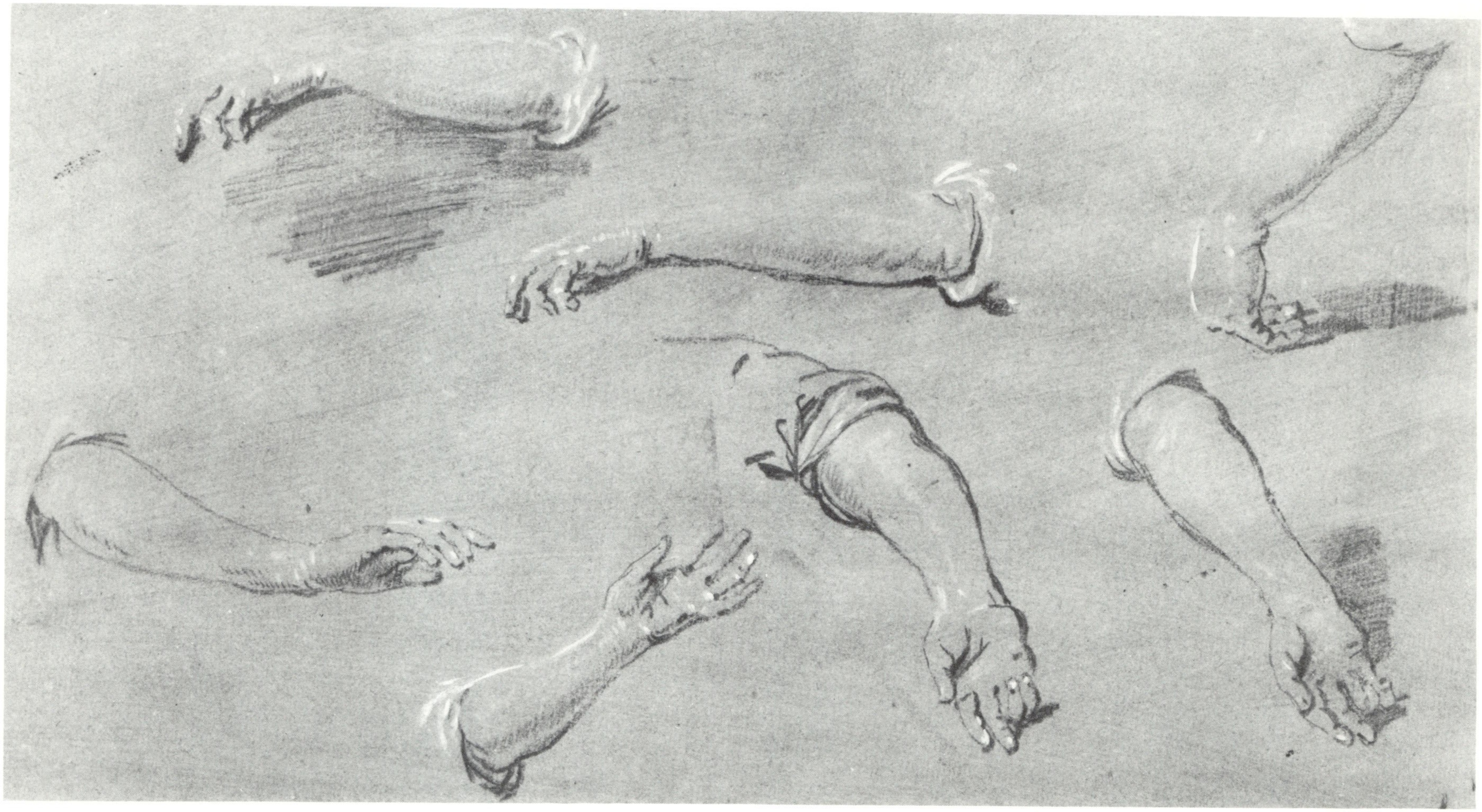

Charcoal and Pencil, heightened with White